Hamlyn all-colour paperbacks

Peter Healey

Microscopes
&Microscopic Life

illustrated by John Bavosi

Hamlyn · London
Sun Books · Melbourne

FOREWORD

This introduction to the microscope and its use explains how the microscope works, how to get the best performance from it, where to find and how to prepare specimens for examination, what equipment is necessary and how to choose it, and many other aspects of the fascinating hobby of microscopy. It then goes on to describe more advanced instruments and techniques, some of which are within reach of the amateur while others are confined to wealthy laboratories.

Microscopes are no longer the tools of a few research scientists; they are found in almost all walks of life, in schools, universities, factories, hospitals and homes. This guide is mainly about microscopes used in biology and about methods of preparing specimens for microscopic examination. Many of the techniques described in this book can easily be carried out by the amateur microscopist at home with simple equipment. As progress is made into more complex methods, the difficulty of obtaining material and cost of the instrumentation will impose their own limitations. It is thus left to the individual to decide how far he or she will progress into the world of microscopic life.

The latter half of the book concerns methods and instruments used in laboratories carrying out research as well as routine procedures. A small excursion is made into the disciplines of microbiology (the study of bacteria), haematology (the study of blood) and histology (the study of tissues), as well as the more usual fields of microscopy such as pond life, marine life and botany. In a world of complex technological advance the light microscope remains virtually as it was fifty years ago and there are still many fresh discoveries for the amateur microscopist.

Published by The Hamlyn Publishing Group Ltd
London · New York · Sydney · Toronto
Hamlyn House, Feltham, Middlesex, England
In association with Sun Books Pty, Ltd. Melbourne.

Copyright © 1969 by The Hamlyn Publishing Group Ltd

SBN 600001032
Phototypeset by BAS Printers Limited, Wallop, Hampshire
Colour separations by Schwitter Limited, Zurich
Printed in England by Sir Joseph Causton & Sons Limited

CONTENTS

HISTORY OF THE MICROSCOPE

Early history

From the very beginning of history, man has explored methods of magnifying all manner of objects to try to find out a little more about them. But he did not progress very far until he discovered some of the properties of glass lenses. Two thousand years ago, the Romans knew that a glass sphere could converge the sun's rays. But it was not until the late 1500s that people became aware of the magnifying power of lenses and used them to study plants and animals.

During the 1600s, several people built simple microscopes. Some engravings of the period show microscopes, but they are obviously inaccurate, particularly as regards size. They show microscopes bigger than the operator! Three prominent Dutchmen in these early days of microscopes were Anton van Leeuwenhoek and Hans and Zacharias Janssen, who were spectacle manufacturers. Leeuwenhoek was one of the first to record observations under a microscope. He made detailed notes and sketches of many of the tiny animals in pond water.

The museum of Middleburg in the Netherlands contains one of the earliest microscopes known, possibly manufactured by one of the Janssens. These early microscopes were extremely simple in construction. They consisted of only two lenses held in position in two sliding tubes. Magnification and focusing were adjusted by moving the tube in or out. Also, only opaque objects could be examined. Then, in the late 1600s, an Italian instrument maker named Campani constructed a microscope for observing transparent specimens.

The images obtained in these early microscopes were dreadfully distorted. The use of oil lamps for illuminating the specimens made matters even worse. In England, the scientist Robert Hooke tried to make better lenses but his results were disappointing. Even so, his observations were important in establishing microscopy as a science.

In the 1700s John Marshall and other microscope manufacturers made great advances in the mechanical design, but they did little to improve the lenses. In the 1800s, however, the optical systems were greatly improved and the greatest advances in the science of microscopy were made.

An early microscopist

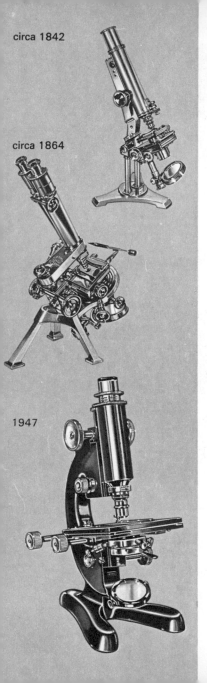

circa 1842

circa 1864

1947

Development of the microscope

Before the 1800s, lens manufacturers had been unable to make lenses that did not split up light into its component colours. This phenomenon, known as *chromatic aberration,* always resulted in a coloured, fuzzy image. If one colour was in focus, the others were not. An English barrister named Chester More Hall had produced corrected lenses for telescopes early in the 1700s, but it was not until about 1830 that suitable corrected microscope lenses, called *achromatic* lenses, were produced. Another common lens defect, *spherical aberration,* also resulted in a blurred image. In 1886, Ernst Abbe, working with Carl Zeiss at Jena, in Germany, produced lenses called *apochromatic* lenses, which were corrected for both chromatic and spherical aberration. The performance of microscopes using these lenses compares favourably with that of modern instruments.

By the late 1800s, microscopes began to look much as they do today, and the large compound versions required manufacturers to seek better illumination systems. In 1893, August Köhler introduced his illumination system, the principle of which is still used today.

Since 1900, microscopes have altered little in principle but very considerably in detail. Such refinements include the introduction of a number of objectives, each with a different magnification, screwed into a revolving turret. Another important development was the microscope stand with a fixed, inclined tube and controls mounted below the stage. One advantage of this system is that the stage remains permanently horizontal. In 1935, Fritz Zernike invented the phase-contrast technique, which makes otherwise invisible specimens visible in a microscope. The latest microscope developments are concerned with the use of built-in illumination and an example of this type of design is the Vickers 'Patholux', with its very solid, square construction and its *ergonomic* (low-placed, easy-to-reach) controls.

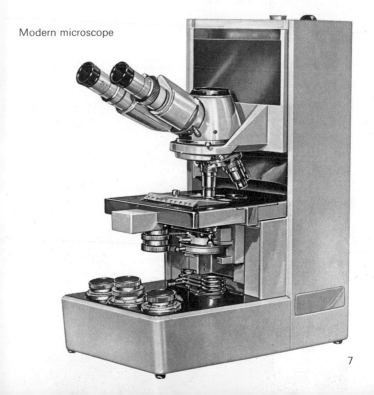

Modern microscope

Microscopes of today

Today, light microscopes have widely varied uses in hospitals, factories, research organizations, and education. In hospitals, the pathologist uses the microscope to aid the diagnosis and cure of disease. Even the surgeon uses a special microscope to enable him to perform delicate operations on the ear and eye. In the electronics industry, the precision assembly of transistors and integrated circuits can be done only with the aid of a microscope. In the assembly shops for such items, hundreds of microscopes are used.

The microscope is without doubt the most important item in any laboratory. It is applied to the study of such diverse subjects as botany and metallurgy. In schools and universities too, study of the microscope and its use are essential parts of scientific education.

The modern light microscope has almost reached the limit of its optical performance. But nevertheless, continual improvements in techniques have led to an ever-growing diversity of uses. As its use spreads, more and more of us need to know at least a little about the way it functions.

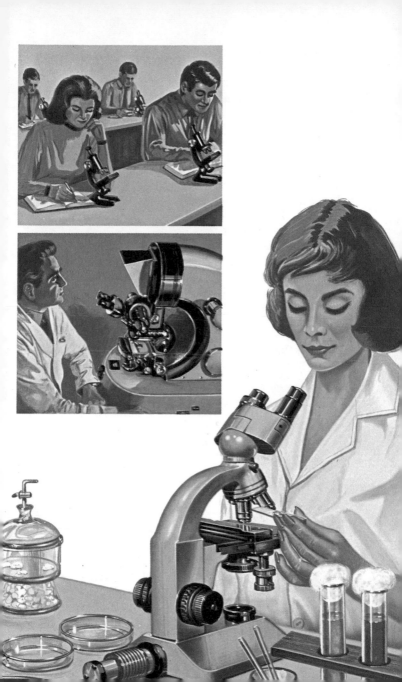

LIGHT AND THE FORMATION OF IMAGES

What is light?

To understand how a microscope works, we must first learn a little optical theory. The first question is *what is light?* Many years ago light was described as a 'vibration in the ether' and even today it can be described best as a form of energy. One of the early theories was the 'bullets' or corpuscular theory. This suggested that light travels in short, consecutive bursts, rather like bullets from a machine gun. But this was soon discovered to be inaccurate.

Huygens' wave theory

In 1678, the Dutch scientist Christiaan Huygens put forward his wave theory, now generally accepted. It states that light travels in the form of a sine-wave, as illustrated. The brightness of the light is proportional to the height or *amplitude* of the wave, and the colour of the light is related to the length of the wave, or wavelength. The wavelength is measured in microns (μ), the usual unit of length in microscopy and there are 1,000 microns in a millimetre.

Blue light has a shorter wavelength than red light.

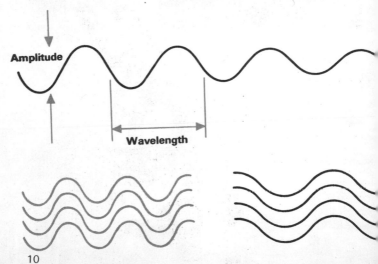

Nature of white light

We know from our observations of rainbows and the spectrum produced when light passes through a prism, that white light is made up of a mixture of colours. The colours of the visible spectrum are violet, indigo, blue, green, yellow, orange and red. Each colour corresponds to a particular wavelength. Violet light has the shortest wavelength in the visible spectrum and red light the longest. Blue light has a wavelength of approximately 0·4 microns, and red light has a wavelength of 0·7 microns. White light consists, therefore, of many rays of light all vibrating at different wavelengths.

Monochromatic light

Light of one wavelength only is called monochromatic light. By definition, it must always be coloured. Although we almost always use microscopes in white light, most modern lenses are designed for use in monochromatic green light. We can improve the resolution by incorporating a blue or green filter in the illumination system. This, of course, only applies to the best modern lenses. For general use, however, filters are neither useful nor desirable.

A prism can split white light into its component colours.

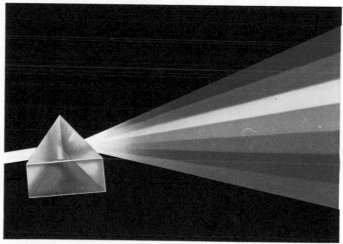

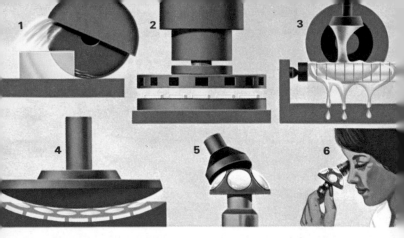

Stages in manufacturing a lens. 1 Cutting glass. 2 Planing. 3 Circular grinding. 4 Lens grinding. 5 Lens polishing. 6 Inspection.

Lenses

The most important components of the microscope are the lenses. They are made from glass or any other transparent material. Face on, most lenses are circular in shape, but in section they vary greatly. They may be convex on both surfaces, concave on both surfaces, crescent-shaped, flat on one side, or almost any combination of these shapes.

Positive and negative lenses

Lenses are divided into two distinct types. Those that cause light rays to converge, that is, to concentrate to form a real image, are called *positive* lenses. Those that cause light rays to diverge, forming no real image, are called *negative* lenses. It is easy to distinguish positive from negative lenses because positive ones are thicker in the middle than at the edges, and vice versa with negative ones.

Magnifying lenses are positive lenses, usually double-convex (convex on both sides) or plano-convex (flat on one side and convex on the other). They magnify by an amount proportional to the curvature of the lens surface. Maximum magnification comes from a spherical lens. The magnifying power of the spherical lens is proportional to its diameter – the smaller the sphere, the greater the magnification. The bottom lens of a highly magnifying, oil-immersion microscope objective is virtually half a sphere – a plano-convex lens only a few tenths of an inch in diameter.

Making lenses

Many earlier lenses were produced by heating and melting a thin glass rod to form a droplet. Even today some are made by melting glass and forming the lens-shape in a mould. Such lenses are only used where correction of defects is not important, such as in the condenser of a microscope lamp.

Most modern microscope lenses are ground from pieces of glass with carborundum or emery powder. The final polishing is done with jeweller's rouge. The manufacture of microscope lenses is a slow and expensive process, and, even though much is now mechanized, careful checking is needed to ensure that the lenses perform in an identical manner. The complex compound lenses are made by cementing a number of accurately ground lenses together with special adhesives.

| Convex | Concave | Plano-concave | Plano-convex |

Light and lenses

The effect that a lens has on rays of light varies with the shape of the lens and also with the density of the substance from which it is made. When a ray of light passes through glass, it is slowed down, or retarded, in proportion to the density of the glass – the greater the density, the greater is the degree of retardation. Rays of light that strike plane glass (that is, glass with its surfaces parallel) at right angles are slowed down but not bent by the glass.

Refraction

If, however, a ray of light strikes a curved glass surface, it is not only slowed down but also deviated. This bending of light rays is called *refraction*. It is caused, as can be seen from the diagrams, by one part of the light ray striking the glass and being slowed down in advance of the adjacent

The speed of light is retarded as it passes through glass. Thus light striking the curved surface of a lens is refracted.

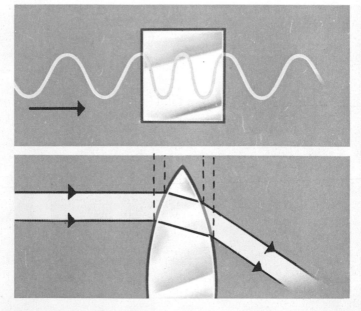

part of the light ray, which continues at its original speed until it also strikes the glass. The deviation is again exaggerated when the rays, travelling at common speed through the glass, emerge from the other side of the lens.

The refraction of light therefore depends on three factors – the angle at which the light ray strikes the lens surface (called the *angle of incidence*), the degree of curvature of the surface, and the density of the glass. By controlling all three factors, the microscope manufacturer can control the optical characteristics, or optics, of the microscope in any manner he chooses.

Refractive index

A figure called the *refractive index* gives a measure of the extent of refraction in glass or any other transparent medium. The refractive index of, say, glass is equal to the ratio of the

Convex lenses cause parallel rays of light to converge. Concave lenses cause them to diverge.

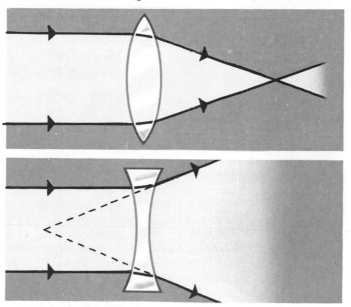

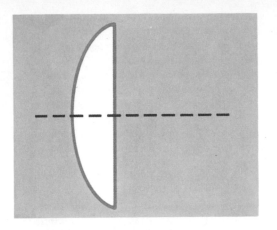

(*Left*) A line drawn at right angles to the vertical axis of a lens is called the *normal*. (*Centre*) Light passing from glass to air is bent away from

sine of the angle of incidence to the sine of the angle of refraction (the angle at which the light ray leaves the air-glass interface). It also equals the ratio of the velocity of light in air to the velocity of light in the substance it passes into, usually glass in the case of lenses. The refractive index of soda glass is 1·53, and of cedarwood oil it is 1·51. The refractive index of air, of course, is 1·0.

Angles of incidence
A line drawn through a lens at right angles to its vertical axis, as illustrated, is called the *normal*. A useful rule to remember is that light passing from air into a more dense medium – for example, a lens – is bent towards the normal, and light passing from a dense medium into air bends away from the normal. This rule applies to light passing into or out of any more dense medium. It explains why a straight stick pushed into water at an angle appears to bend at a point level with the surface, and why an object under water appears to be closer to the surface than it really is.

When the angle at which light strikes a lens is increased, the rays emerging from the back of the lens bend further away from the normal. That is, they get closer and closer to

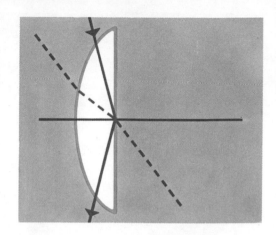

the normal. (*Right*) Beyond a certain angle of incidence, a state of total internal reflection is reached.

the surface of the glass. At a certain angle of incidence, the rays emerge from the back of the lens parallel to the surface. If the angle of incidence is increased beyond this point, no rays emerge at all. The light is reflected internally within the lens, producing a state of *total internal reflection*. We see therefore, that although light can always pass into one side of a lens, it will not necessarily always emerge on the opposite side. If the angle of incidence is large enough, it will even emerge on the same side as it entered, so that the lens acts as a mirror.

It is a simple task to demonstrate this phenomenon. Take a lens or even a piece of plane glass, hold it up so that you can see through it, and slowly rotate it about its vertical axis. Suddenly you will find that you can no longer see through the glass but can see instead the reflection of some object to one side of you. What has happened is that the angle of incidence has been slowly increased until a state of total internal reflection has been reached. The piece of glass then acts as a mirror held up at an angle.

By understanding these few basic optical facts, we now have sufficient knowledge at our disposal to begin to see how images are formed.

17

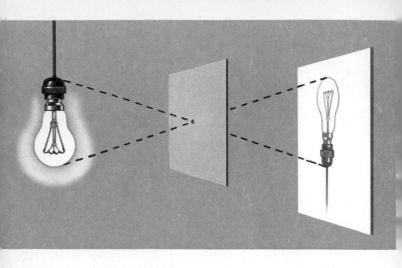

The pin-hole camera

How images are formed
Forming an image with a pin-hole camera

First we must look at the simplest way of forming an image. Take a piece of card with a tiny hole in the centre, and place it between a bright object such as a light bulb, and a screen. You will find that an image of the lamp is formed on the screen. You have constructed the simplest form of imaging device – a *pin-hole camera*. You will see immediately that the image is very dim and is upside-down, or inverted. By enlarging the hole, you do not increase the image brightness to any great extent, but merely make the image more blurred.

From this evidence it is easy to see what is happening. Rays of light are being emitted by the lamp in straight lines in all directions and only the few rays of light that pass through the pin-hole form the image. That is why the image is so dull. A perfect image would be formed if the hole were so small that only a single ray of light from a single point on the lamp could pass through, but it is doubtful whether it could be seen. But, in fact, several rays from the same point on the lamp pass through the hole and form a circular patch on the screen instead of a single point. The area of the patch

Enlarging the hole blurs the image (*above*) because of the spread of a single point (*below*) from the source over a large area on the screen.

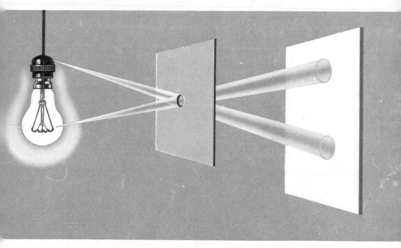

is proportional to the size of the hole. Rays from an adjacent point on the lamp will form a patch in a similar manner, and overlap the other patch to form a blurred image – the bigger the hole, the more blurred the image becomes.

Forming an image with a lens

Real image We have already seen that lenses have the property of bending, or refracting, light. By placing a lens in the hole of a pin-hole camera, we can overcome the problem of light travelling only in straight lines. The lens can gather all the rays of light emanating from a single point on the object and bring them together again to form a single point on the screen. The rays of light from adjacent points on the object will form corresponding points on the screen, and, as shown in the illustrations at the bottom of these two pages, a bright, sharp, inverted *real* image will result.

This will only work, of course, if the screen is in the focus position. Any movement away from this position will produce an out-of-focus picture because the rays of light from the lens have either not yet come together to form a focus, or have crossed over beyond the focus to produce large, overlapping areas of light, as in the case of the pin-hole camera. The further the screen is moved away from the focus position, the more blurred and dim the image becomes. The brightness is also decreased because the same amount of light has been spread over a larger area. This is demonstrated in the illustration on the right.

If a positive lens is inserted, a bright image can be formed.

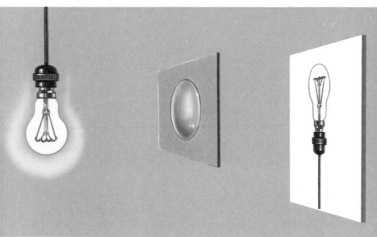

One basic characteristic of a simple lens is its *principal focus*. This is the point at which parallel rays of light entering the lens are brought together. The distance therefore between the principal focus and the lens is called the *focal length*. ·

In addition to its principal focus, a lens has other pairs

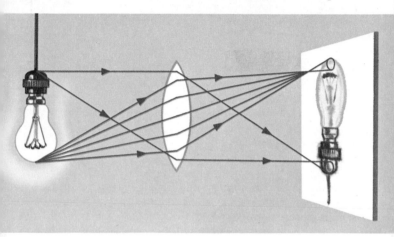

If the screen is not in the focus position, a blurred image will result.

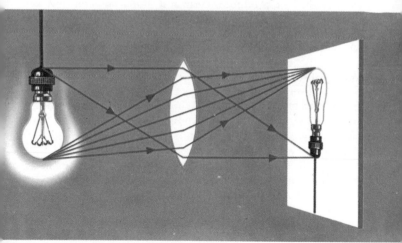

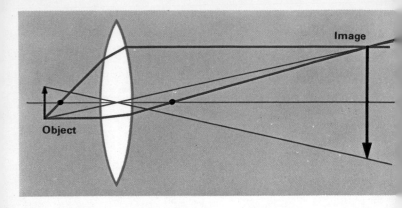

Location of conjugate foci

of *conjugate foci*, on either side of the lens. An object placed at one conjugate focus will produce a clear image on a screen placed at the other. If the object is a long way from the lens, the image produced will be small and quite close to the lens. If we now reverse the position of the object and the screen so that the object is close to the lens and the screen is far away, we produce a clear and much larger image. We see, therefore, that the closer the lens is to the object, the larger and further away is the image and this is demonstrated in the top two illustrations on the right. But this only holds true if the object is further away from the lens than the principal focus.

Virtual image When the object is placed at the principal focus, the image is formed at infinity. Any further movement in the same direction produces an image on the same side as the object. This image is called a *virtual image,* as opposed to a real image, because it cannot be focused on a screen. It can be viewed only by looking through the lens. As shown in the lower illustration on the opposite page, it is upright and not inverted like the other images we have examined. This is the kind of image we see when we look through a magnifying glass, which is, in a sense, the simplest of all microscopes.

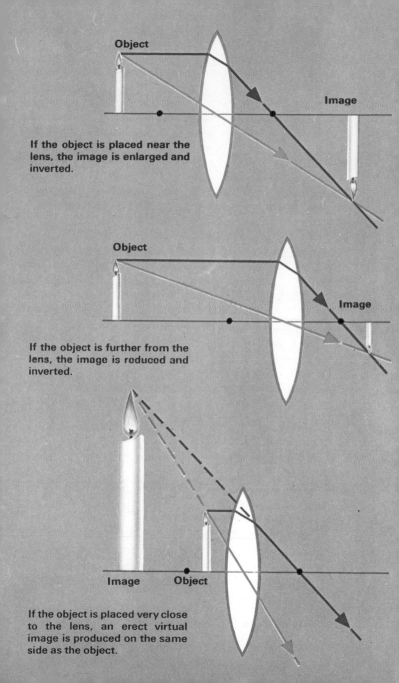

Object

Image

If the object is placed near the lens, the image is enlarged and inverted.

Object

Image

If the object is further from the lens, the image is reduced and inverted.

Image   **Object**

If the object is placed very close to the lens, an erect virtual image is produced on the same side as the object.

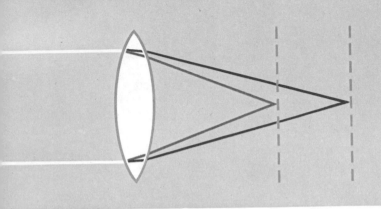

Chromatic aberration in a simple lens

Lens defects
Chromatic aberration

Unfortunately the formation of images in the manner described only works in theory, and the images formed in practice are not of a quality suitable for microscopy. This is partly because most of the rules that we have applied only hold true for single rays of monochromatic light. In microscopy, of course, we have to use many rays of white light, which is composed of a mixture of colours. The problem is that a lens does not refract all the colours to the same degree – the shorter the wavelength of the light, the greater is the degree of bending. Violet light, for example, which has a short wavelength, is refracted to a greater degree than red light which has a long wavelength.

Because of this difference in refraction, the lens splits the white light into its component parts as it passes through it. This phenomenon was first noticed by Isaac Newton in 1666 when he shone white light through a triangular prism and observed that a spectrum was produced on a screen placed beyond the prism. A similar thing happens when we shine a beam of white light through a lens on to a screen. Circles of coloured light are produced. If the screen is placed close to the lens, there is a patch of blue light surrounded by red. If the screen is placed further away, this changes to red light surrounded by blue. There is no position in which the object from which the light is emanating is in sharp focus without colour interference. This serious defect is called *chromatic aberration.*

It took the early lens makers some sixty years (1770–1830) to correct this aberration, after which microscopy became a respectable science.

Correcting chromatic aberration When Isaac Newton performed his original experiments with prisms, he failed to observe that different types of glass have different effects

Chromatic aberration can be corrected by using light of one colour.

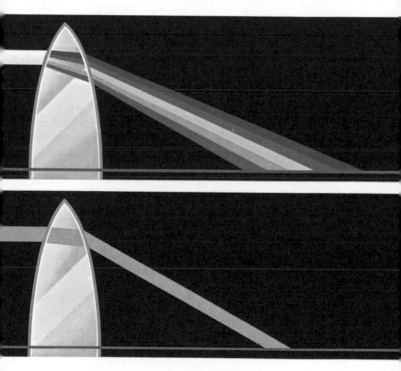

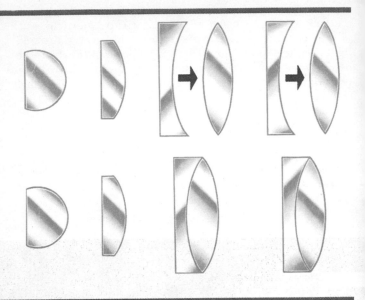

The lenses of an achromatic objective

on the degree of refraction of the various colours. When this fact was discovered, it became apparent that chromatic aberration could be cured to some extent by making lenses from combinations of different glasses.

The simple method of correcting chromatic aberration by combining two different types of glass into a compound lens does not, unfortunately, give perfect results. The manufacturers of microscope lenses compromise by making the major correction for green light and combining the other colours in pairs and treating them as one. This system gives quite acceptable results and the lenses thus produced are called *achromatic*.

The major defect of the achromatic lens is that very small objects, which may be colourless, may appear when viewed down the microscope to be greenish in colour. To overcome such problems, the mineral fluorite is incorporated into the design of the objective. By this means, correction is made not

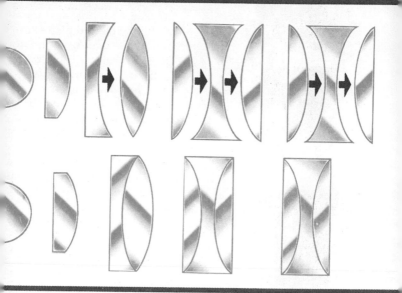

The lenses of an apochromatic objective

for two, but for three colours. These *apochromatic* objectives, as they are called, were first produced in 1886 by Ernst Abbe, working for Carl Zeiss. They gave results far superior in brightness and resolution, as well as in colour correction, to the lenses then being made.

The apochromatic lens is extremely complex in design. Full correction is obtained by a combination of the objective lens and the eyepiece. Apochromatic objectives must, therefore, be used in conjunction with a corresponding compensating eyepiece, which is used to counter-balance size differences in the coloured parts of the image. Such complexity is, of course, extremely costly, and apochromatic lenses are only normally used where the ultimate performance is imperative or where good quality micro-photography is required.

If achromatic lenses are used for micro-photography, coloured filters must be used because photographic emulsions

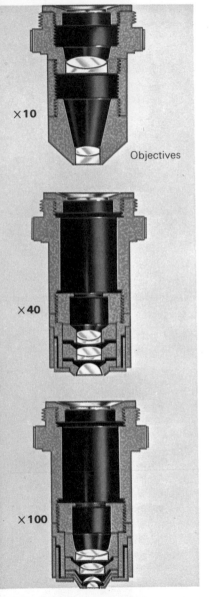

×10

×40

×100

Objectives

are usually more sensitive to blue light than to green. A slightly blurred picture can result if the image contains several colours which come to a focus in different planes. Apochromatic objectives, on the other hand, can give sharp micrographs even without the use of colour filters.

Spherical aberration

The other major defect that lenses suffer from is called *spherical aberration*. It is important not to confuse this with a completely different problem called *field curvature,* in which the edges of the observed image are not in focus at the same time as the central part.

To understand spherical aberration, we must realize that the rays of light entering the microscope objective come not from a source in air, but from glass – the microscope slide or coverslip. Rays of light are brought to a cone by the microscope condenser lens and pass through the slide. The rays at the edges of the cone are refracted to a different degree from the rays in the middle. In fact, if these rays were traced back, the outer rays would appear to come from a source higher in the slide than the inner ones. For this reason the light rays entering the objective lens are not formed into a perfect cone, and the lens is unable to bring them to a perfect focus. The result is that the lens focuses several images at once, each one superimposed on the other.

Correcting spherical aberration The extent of the aberration varies with the thickness of the glass from which the cone of light is emerging. The lens designer has to take this into account when computing his lens curves in order to reduce the aberration as much as possible. In practice, only high-magnification lenses cause trouble in this respect. The problem is overcome in the highest powers by joining the bottom of the objective lens to the slide with an oil of the same refractive index as glass.

Spherical aberration. Rays passing through the edge of a lens come to focus in a different place from those passing near the centre.

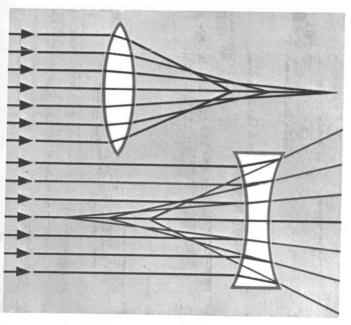

THE COMPOUND MICROSCOPE

General principles

The simple microscope, as we have seen, consists of a single lens system and produces a magnified image. But there is a limit to the magnification that can be obtained in this way. The compound microscope was developed to give greater magnification than the simple microscope. It consists of two main lens systems. A magnified image produced by the first lens system, is viewed and further enlarged by the second. The primary magnification, by the *objective* lens, controls both the major magnification and the quality, the image falling in such a position that it can be viewed by the second lens system – the *eyepiece*.

The specimen to be examined is illuminated by another lens system called the *condenser*. Light, directed into the bottom of the condenser by the plane side of a *mirror,* is formed into a cone and focused on an area of the specimen that is at least as big as the area covered by the objective, so as to fill the field of view with light.

If we take two rays of light being emitted from the edge of the field, we find that one travels up the microscope parallel to the optical axis until it reaches the objective lens, where it is refracted towards the normal. The other passes through the lens centre and is not refracted. Both rays come to a focus further up the body-tube of the microscope. All the rays passing through the microscope can be considered to behave in a similar way and a real, but reversed, image is formed. This is the primary aerial image that is examined by the eyepiece.

If we now place another lens slightly above this image, the rays will have crossed over before they enter the lens, and a similar situation to the first develops. One ray passes through the centre of the lens, and one passes through the edge and is refracted. But now we find that they do not intersect but actually diverge. If we now trace them backwards, they eventually come to a focus to form an erect virtual image of the primary image. This second image is the one the eye sees. It is this two-stage magnification that is the principle of the compound light microscope.

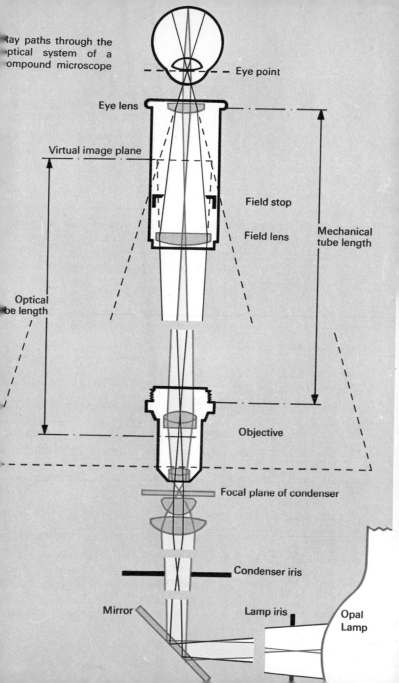

Ray paths through the optical system of a compound microscope

Eye point

Eye lens

Virtual image plane

Field stop

Field lens

Mechanical tube length

Optical tube length

Objective

Focal plane of condenser

Condenser iris

Mirror

Lamp iris

Opal Lamp

The objective lens

The objective lens is the most important lens system of the microscope, controlling both the magnification available and the image quality. Most microscopes have three or four objective lenses screwed into a revolving turret at the bottom of the body-tube. Ideally, when a new objective is brought into use, it should be in exactly the same position in relation to the field of view as the previous lens and also near to the exact focus. Objectives that measure up to these conditions are said to be both *par-focal* and *par-central*.

All objectives made today are fitted to the microscope by means of a standard screw-thread, so that they can be moved from microscope to microscope regardless of the manufacturer. Such changes, however, often mean that standards of par-focality and par-centricity are lost. It is rarely possible to use a combination of objectives by different manufacturers in the same turret.

Microscope objectives

The objectives screw into the turret.

Magnification

Objective lenses with a great variety of magnifications are available – from ×1 to over ×100. The ones most widely used are ×5, ×10, ×20, ×40 and ×100. The ×100 lenses are used in conjunction with immersion oil which connects the bottom of the objective to the specimen slide.

Lens markings

If we examine an objective lens, we see figures engraved on the side. One of them gives the magnification or, if the lens is an old one, the focal length in inches – for example, $\frac{2}{3}$, $\frac{1}{6}$, and $\frac{1}{12}$. The $\frac{2}{3}$ inch lens is approximately equivalent to a ×10 objective, the $\frac{1}{6}$ inch to a ×40, and the $\frac{1}{12}$ inch to a ×100 (oil-immersion). Another figure engraved on the side gives the *numerical aperture* (N.A.) of the lens, which is a most important guide to performance (see page 39). Lenses used without immersion oil have a numerical aperture up to a maximum of 1·0. Oil-immersion lenses have a maximum

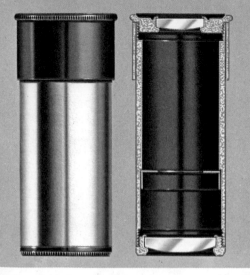

A Huygenian eyepiece   A compensating eyepiece

of 1·5. Other figures – for example, 160/0·17 – will also be found on some lenses. The 160 refers to the tube length (in mm.) with which the lens is designed to be used and the 0·17 refers to the thickness of the cover-slip (normal No. 1 type) that should be used.

Many modern objectives are long enough to be racked into the specimen slide if the coarse focusing control is turned too far. Most of them, therefore, have bottom lens components spring-loaded to protect both the lens and the slides.

The eyepiece

The microscope eyepiece, or *ocular*, is used to examine and magnify the image produced by the objective lens. For normal purposes we can consider two types of eyepieces – the *Huygenian* eyepiece, used in conjunction with ordinary achromatic objectives, and the *compensating* eyepiece, used with apochromatic objectives (see page 26).

The Huygenian eyepiece is composed simply of two lenses. The bottom lens collects the image from the objective, reduces it, and re-forms it within the eyepiece at the level of the *field stop*. The field stop can be seen by unscrewing the top lens of

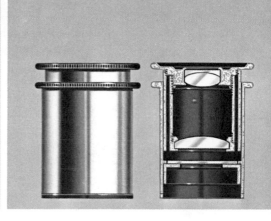

A Kellner eyepiece with graticule

the eyepiece. Any object placed at the level of the field stop will be precisely in focus in the field of view. The upper eyepice lens forms an enlarged virtual image for viewing.

The cross-over point of the rays of light as they emerge from the top lens of the eyepiece is actually a small disc of light, called the *eyepoint,* or *exit pupil.* This is where the eye is placed to view the image. The size and position of the eyepoint will vary from instrument to instrument. The size will also be controlled by the degree of magnification of the microscope. Wearers of spectacles should therefore use eyepieces with a high eyepoint position so that they can keep on their spectacles when looking through the microscope.

Magnification

Eyepieces normally range in magnification from × 5 to × 15, although most microscopists tend to use only one eyepiece, relying on changing objectives to alter magnification. Eyepieces with magnification above × 10 are best employed only with very high-class objectives, because the image quality produced by the lower-quality objectives is not sufficiently good to stand up to the high secondary magnification.

Magnifying power

This is calculated by multiplying the magnification of the objective by that of the eyepiece, provided that the tube length is correct (usually 160 mm.), and there is no intermediate lens or binocular-head magnification to take into account. Adjustable tube-lengths should always be set at 160 mm. Any objective will be working at its optimum if the total magnification employed is 1,000 times the numerical aperture of the lens. For example, if the objective lens is ×100 and its aperture is 1·2, then a ×12 eyepiece will give a total magnification of ×1,200.

There is a limit to the magnification of a microscope. In the optical image formed, each point reproduced from the object is, because of diffraction, spread over a small area. Therefore, each point on the object is represented by a corresponding disc in the image. When a certain magnification has been reached, the detail in the object becomes the same size as the disc, and more magnification will not improve it.

If you magnify a blurred image, the blurring is also magnified.

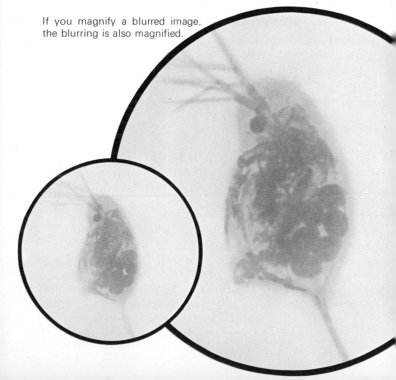

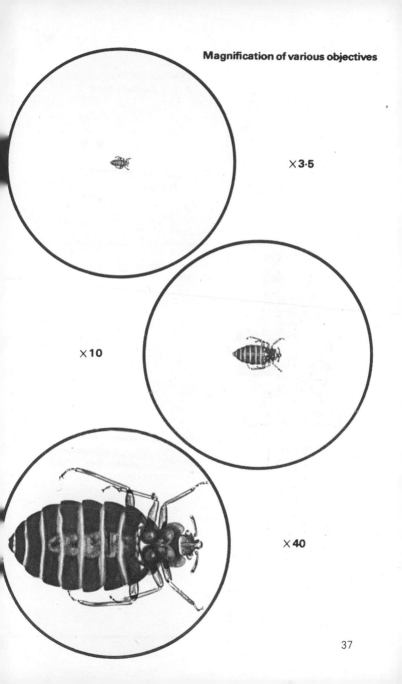

Magnification of various objectives

×3·5

×10

×40

The human eye can separate dots 0·25 mm. apart.

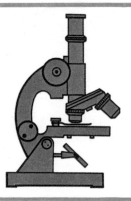

The light microscope can separate dots 0·25 microns apart

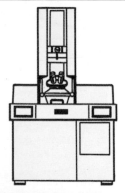

The electron microscope can separate dots less than 5 Angstrom units apart.

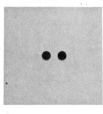

Resolving power

The most important feature of an objective lens is its ability to reproduce detail, that is, its *resolving power*, and this is directly linked to its numerical aperture.

Resolving power is best described as the ability of a lens to separate lines or dots. The resolution of the human eye varies from person to person, but, at our normal working distance of 25 cm., we should be able to resolve 0·25 mm. This means that if two dots on a piece of paper are more than 0·25 mm. apart, we should be able to see the gap between them. If they are closer together, they will blur and appear as one dot. The light microscope can resolve dots that are about 0·25 microns apart, depending on the quality of the objective used. To see this separation, we must use a magnification of × 1,000 to bring the distance of 0·25 microns up to 0·25 mm.

The resolution of the microscope depends on two factors – the wavelength of the light employed and the *numerical aperture* of the objective. The numerical aperture is a measure of the size of the cone of light that the objective will accept. It is expressed not as an angle in degrees but numerically in terms of the sine of the angle. If we examine the light coming up the microscope, we see that its effective source is the specimen. The size of the cone that the objective can accept is, therefore, governed by the design of its bottom lens and its working distance. Also, as can be seen in the illustrations, the amount of refraction that occurs as the light leaves the specimen cover-glass must be taken into account. Numerical aperture is, in fact, the sine of half the angle of acceptance multiplied by the refractive index of the medium through which the light is passing, air or immersion oil (R.I. 1·515).

The highest theoretical numerical aperture for a dry lens system is 1·0 and for an oil-immersion system, 1·5. Many modern objectives have numerical apertures close to these figures and it is necessary to use a condenser lens on the microscope which can provide a sufficiently large cone of light to match the objective. The simple formula below enables us to calculate the resolving power of a given combination of objective and condenser.

$$\frac{\text{Resolution}}{\text{in microns}} = \frac{\text{Wavelength of light in microns}}{\text{N.A. of objective} + \text{N.A. of condenser}}$$

The condenser

The cone of light that illuminates the specimen is provided by the condenser lens system, mounted below the stage. This mounting is normally arranged so that the condenser can be raised or lowered. On the better microscopes, the condenser can also be centred. The condenser lens is used to control the illumination entering the objective lens; it must have a numerical aperture high enough to give the objective the cone of light it requires. It is no use, for example, coupling the cheapest two-lens Abbe condenser with an expensive apochromatic objective. On the other hand, the condenser lens must be able to fill the large field of a low-power objective. These requirements are conflicting because the higher the numerical aperture, the closer is the working distance, and the smaller is the field covered by the illumination spot.

To overcome this problem, some condensers are fitted with a top lens which can be swung out of the system at the turn of a knob. In this way, the power of the condenser can be reduced and the larger fields illuminated. With the top lens in position, the numerical aperture is increased, and high-power objectives can be used without deterioration of performance. Another way of overcoming the problem is to unscrew the top lens, or lenses. The drawback with this method is that it is usually necessary to remove the condenser from the microscope to unscrew the desired lens.

It is incorrect procedure to lower the condenser to fill the field of view. If the condenser is unable to fill the field of the lower-power objectives, the only correct thing to do is either to change it for one that will or to remove it completely. It is for this purpose that the microscope mirror is provided with a concave side. When the condenser is not being used, the concave mirror itself converges the rays of light from the lamp on to the specimen. Of course, whenever the condenser is being used, the plane side of the mirror is the only one that should be employed.

At the bottom of the condenser, there is a variable aperture, or *iris diaphragm*, controlled by a lever at the side, and a filter tray to house coloured or daylight filters. The correct use of the iris diaphragm is of the utmost importance in microscopy, as will be explained in detail later (see page 48).

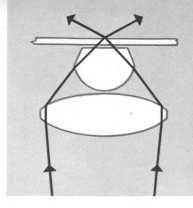

Two-lens Abbe condenser

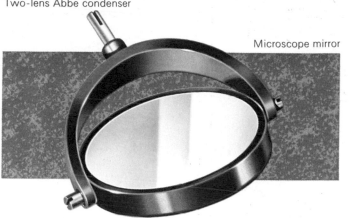

Microscope mirror

An achromatic condenser

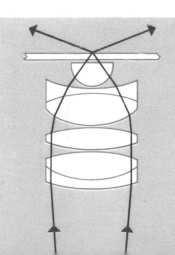

Coarse-focus mechanism

The mechanics of the microscope

Almost all the mechanical parts of the microscope are confined to the stand. They consist of controls for coarse and fine focusing; for raising and lowering the condenser, and for centring the condenser. Some microscopes have a mechanical stage on which the specimen can be tracked accurately in two directions at right angles to each other. Others have a stage called a *goniometer stage*, on which the specimen can be rotated about the optical axis of the microscope. The mirror is normally fitted on a gimbal system so that it can be easily positioned.

The tendency in recent years has been to place the controls low down on the microscope for easy use. Also, it is normal practice with modern instruments to focus by moving the stage up and down rather than by moving the microscope tube itself, but the principle is the same. The condenser and coarse-focusing controls are used to set up the microscope, but once set, they are left until a change of magnification is needed. Normally only the fine-focus control and stage movements are used.

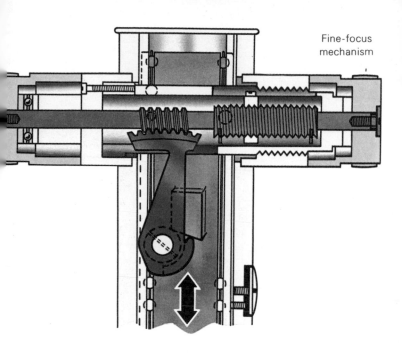

The coarse-focusing and condenser controls are almost invariably operated by rack-and-pinion motion. Turning the control knob rotates the pinion, which, in turn, drives the rack with the moving part attached. These parts must be accurately matched together or two faults may appear – backlash may develop in the system or the weight of the moving part may cause the rack to run down. Neither of these faults can be readily corrected by the amateur microscopist. To guard against them, rack-and-pinion systems should be kept clean and occasionally lubricated with the correct grease which can be obtained from the manufacturer.

Most fine-focus mechanisms consist of a screw-and-lever system. The screw drives one end of the lever, while the other end of the lever, after providing the required reduction in motion, drives the part to be moved. Care must be taken with most fine-focus systems not to jam the screw tight when the end of the motion has been reached. Some microscopes have indicator lines to show the end of the travel.

Mechanical stages, usually driven by rack and pinion, are very useful components to have on the microscope. But care must be taken not to spill immersion oil or mounting media into the mechanism because the movements of the mechanical stage at high magnification need to be very precise.

The lamp
Illumination for the microscope can be provided either by an external light source, directed up the microscope by the mirror, or by a lamp built into the base. The simplest form of illumination is daylight, preferably bright cloud, but this has so many limitations that it is rarely of any practical use.

The simplest practical microscope lamp is an ordinary desk lamp or a bulb mounted in a tin. The bulb should be frosted to diffuse the light and, if fairly high-power work is to be done, should be 60 or 100 watts, as the human eye works best at fairly high levels of illumination.

Our simple microscope lamp can be further improved by adding an iris diaphragm in front of the bulb to control the light output. If the illumination system is to be correctly set up, this diaphragm is essential. Another essential feature of a microscope lamp is that it must be able to fill completely and evenly the back lens of the objective. This can be checked by removing the eyepiece and looking down the tube with the condenser in its uppermost position.

Köhler illumination
The best external microscope illuminator is the high-intensity lamp. The light source is a 6 or 12 volt bulb, driven from a

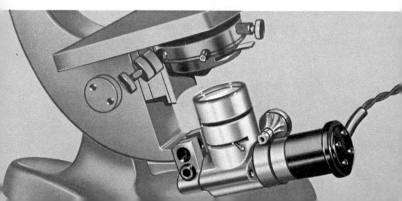

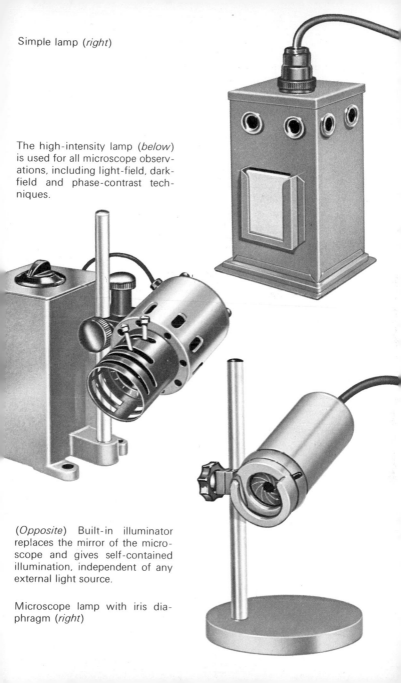

Simple lamp (*right*)

The high-intensity lamp (*below*) is used for all microscope observations, including light-field, dark-field and phase-contrast techniques.

(*Opposite*) Built-in illuminator replaces the mirror of the microscope and gives self-contained illumination, independent of any external light source.

Microscope lamp with iris diaphragm (*right*)

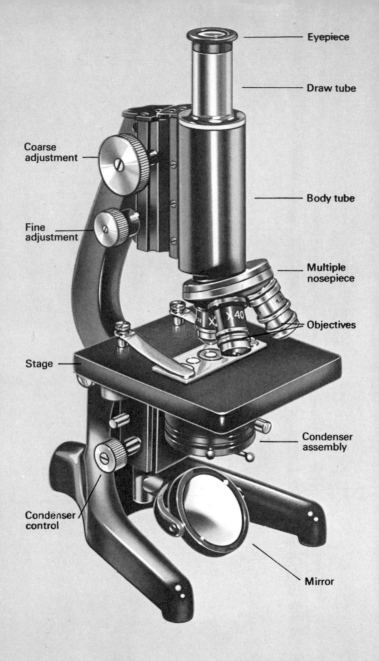

Eyepiece

Draw tube

Coarse adjustment

Fine adjustment

Body tube

Multiple nosepiece

Objectives

Stage

Condenser assembly

Condenser control

Mirror

transformer. A rheostat in the system controls the voltage of the lamp and therefore the light intensity. In front of the bulb is a movable condenser lens incorporating an iris diaphragm. The bulbs used in these lamps have small filaments, usually wound in the form of a square. The bulb-holder is equipped with centring screws, by which the filaments can be aligned to the axis of the condenser. By carefully adjusting the variable controls of the lamp, the illumination can be set up so as to get the best results from the microscope. This system is called Köhler illumination after its inventor, August Köhler.

Many modern microscopes have a built-in illumination system in which a movable external mirror is not necessary. The microscope can therefore be moved around without upsetting the critical mirror setting. As with the external lamp, these built-in lamps can vary considerably in quality, from a simple mains lamp up to a full Köhler illuminator.

Using the microscope

The home microscopist will almost certainly equip himself initially with a microscope with an external light source, probably just a simple lamp with an iris diaphragm. For this reason, our setting-up procedure will describe such a system.

If possible, the microscope should be left permanently set up on the bench or table and covered when not in use with a plastic dust-cover. Generally, most damage is done to the microscope by the constant knocking it gets when being moved in and out of its box. If the microscope cannot be left permanently set up, then it is often a good plan to make some form of base plate to hold the microscope and lamp in their relative positions. In this way, far less setting up is required each time the equipment is brought into use.

The microscope table should be as solid as possible. Any vibration will make any sort of critical microscopy at high magnifications quite impossible. A comfortable sitting position at the microscope is most important and for this reason care should be taken to get the height correct. Remember that many microscope preparations require the stage to be horizontal, and it is therefore not always possible to incline the microscope if this also involves inclining the stage.

Setting up the illumination

Let us now examine how to set up the microscope illumination so that we can get the best from the equipment we have. The microscope lamp should be directed at the mirror and placed about ten inches in front of it. Almost all microscopists have, at some time in their career, discovered something wrong with the illumination, only to find that they are not using the microscope lamp at all as a source of illumination, but some other distant source, such as a room light.

With the microscope mirror illuminated, rack up the condenser until it is in its uppermost position. Then, while looking down the microscope with a low-power objective (about ×10), move the mirror around until the light is at its brightest. Place the specimen slide on the stage and focus it with the coarse-focus control. With the microscope in focus, we can set the condenser height. Shut down the iris diaphragm on the lamp. It should then be possible to see an

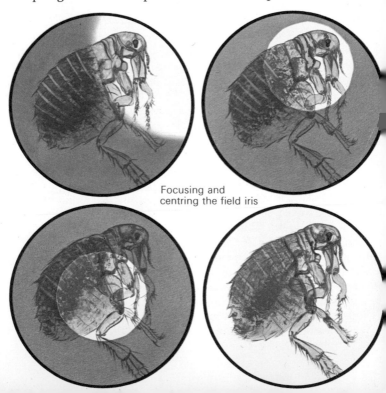

Focusing and
centring the field iris

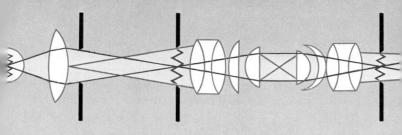

Köhler illumination

out-of-focus image of this iris in the field of view. If the image is off-centre, bring it back by slight adjustment to the mirror. Then focus the iris with the condenser height-adjustment control, and, when it is in sharp focus, open it until it just clears the field of view. With the lower-powered objectives, it is most likely that the field will not be fully filled, even when the diaphragm is fully open. If this happens, it is necessary either to remove a lens from the top of the condenser or to place a piece of ground glass in the condenser filter-tray.

The field iris has now been correctly set, leaving only the setting of the condenser iris before the microscope is ready for use. The condenser iris is in focus at the back focal plane of the objective lens. To examine it, remove the eyepiece and look down the tube. With the condenser iris fully open, the back lens of the objective should be filled with light.

Best results are obtained when about two-thirds of this cone are used, so stop down the iris on the condenser until this condition is reached. If the condenser diaphragm is open too far, glare will result, and if it is stopped down beyond the two-thirds position, there will be a loss of resolution. The correct use of this control is of the utmost importance in obtaining the best results. These conditions will have to be adjusted slightly for each objective used.

The illumination system described above is called *critical illumination*. It is the best we can do with the simple microscope lamp, and is quite satisfactory for most purposes. For the very best results, the Köhler system must be used. For this a microscope lamp with a condenser and an iris in front of it is required. Use the lamp condenser to focus an image of the lamp filament, via the mirror, on to the iris of the microscope condenser. Then continue as for setting up with the simple lamp. Anyone not fortunate enough to have an iris diaphragm in front of his lamp, can overcome the problem of setting the condenser height by placing a pointer in front of, and close to, the lamp and adjusting the condenser until the pointer is in focus in the field of view. The pointer can then be removed, and the microscope is ready for use.

Focusing

There is a set sequence of operations that makes focusing on a specimen easier and quicker and at the same time prevents damage to both specimen and microscope. The easiest and best objective to use for initial focusing is the × 10, because there is a stop on most microscopes to prevent this lens being racked down into the microslide. Most other objectives of higher magnification can be racked right down. Place the slide on the stage and rack down the body-tube until either the stop is reached or the objective is close to, but never touching, the cover-slip. Then, using the coarse control, rack up the body-tube until the specimen comes into focus. Do not rack down while looking down the microscope because if there is no stop damage may result.

Oil-immersion technique

Using the oil-immersion lens requires a somewhat different technique. Place the specimen slide on the stage with the body-tube racked fairly high. With the aid of some form of applicator, such as a glass rod, put a small drop of immersion oil on the slide immediately below the objective. Select the oil immersion lens, if this has not already been done, and rack down until the bottom lens just makes contact with the oil. Look down the microscope and gently rack down with the coarse adjustment until the image begins to come into focus. Focus finally with the fine-focus control.

This is the best technique for the practised operator, who will soon be able to feel when the point of focus has been passed and there is danger of contact with the slide. The oil film tends to act as a cushion, but even so, only the gentlest pressure should be exerted on the focus control. Always wipe the objective with soft tissue immediately after use.

1 Place slide on the stage.

2 Rack down while watching.

3 Rack up to focus.

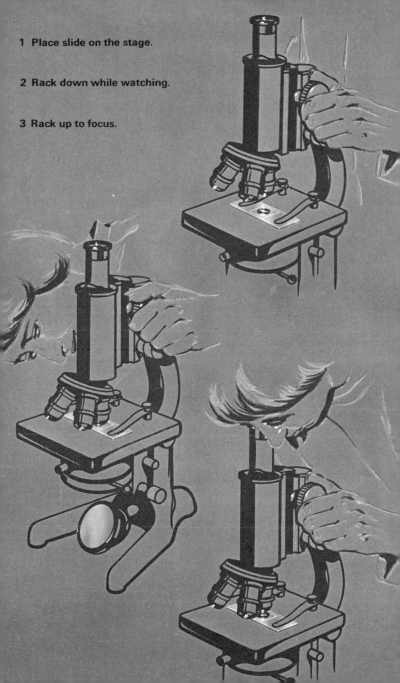

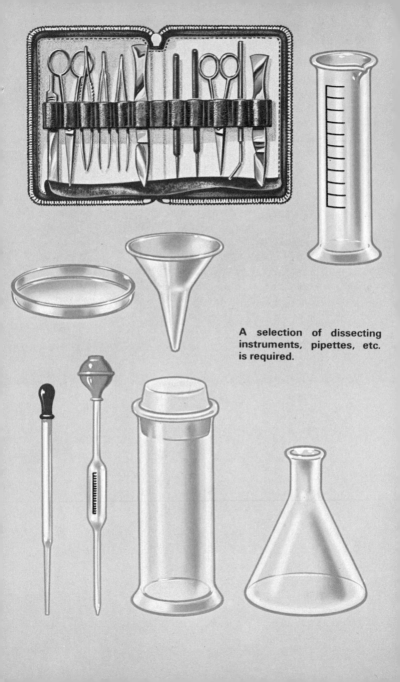

A selection of dissecting
instruments, pipettes, etc.
is required.

MICROSCOPY FOR THE AMATEUR

Basic equipment

One of the first problems that presents itself to the amateur is that of selecting the basic equipment necessary to start microscopy without involving too great an outlay. The choice of the microscope will be governed by the amount of money available, and there is no doubt that the most economical method is to purchase a second-hand instrument from a reputable dealer. A full-sized microscope should be purchased if possible because, with one notable exception (the Cooke-McArthur microscope), the smaller microscopes available today fall short of the desired performance in many respects.

There are snags to buying second-hand equipment, but it is easy to check the condition of a microscope. All the controls should be smooth but with that taut feeling of well-matched mechanical components. Lenses should be examined to ensure they are free from chips, scratches, or blemishes. It is always a good sign if they are stored neatly away in the appropriate containers. Age is of little importance; microscopes and lenses properly looked after will last indefinitely.

In microscopy the simplest set-up will often give the best results. A simple, straight monocular stand with a good eyepiece and objective cannot be improved upon from the point of view of optical performance and ease of use.

The final choice must depend on personal preference, but a most satisfactory selection would include a simple stand with mirror and condenser-height control, a two-lens Abbe condenser, three objectives of powers $\times 2 \cdot 5$, $\times 10$ and $\times 40$, and eyepieces of $\times 5$ and $\times 10$. Other objectives, a rotating nose-piece, a mechanical stage, and so on, can be added later. The microscope lamp can be bought or easily made at home. An ordinary desk lamp will do if no other is available.

A selection of pipettes, slides, dishes, jars, and so on, will be needed. Even some advanced research departments use old jam jars for specimen storage. A few dissecting instruments, such as scissors, forceps, and dissecting needles, will also be needed, but it is false economy to buy cheap ones. Cheap instruments will rust and quickly become useless, whereas those made from stainless steel will last for many years.

Preparing specimens

It is essential in microscopy to have a well-prepared specimen to look at. Simple specimen preparation is not difficult and requires only a few simple tools, but it requires a lot of practice. For almost any specimen preparation, slides and cover-slips will be required. Most micro-slides are pieces of plane glass about 1 mm. thick, measuring 3 inches by 1 inch. The thickness is most important. For most purposes, slides of 1 mm. to 1·2 mm. can be used, but with apochromatic objectives and a high-powered condenser, slides of 0·8 mm. to 1 mm. should be used. Other sizes of slides are also available.

Cover-slips vary greatly in size, but the most common ones are $\frac{7}{8}$ inch square. Again, the thickness is important and different thicknesses are available. Generally, cover-slips are designated by a single number. The most common and the best one for most purposes is the No. 1 cover-slip, which is about 0·17 mm. in thickness.

Both slides and cover-slips must be cleaned before use.

Slides and cover-slips

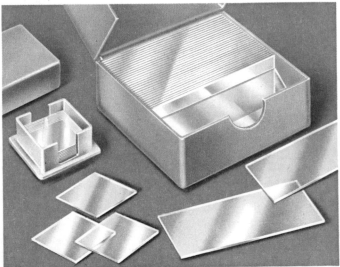

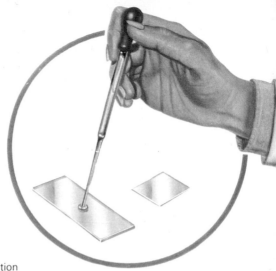

Making a wet preparation

The ideal way is to store them in spirit and polish them immediately before use with a clean, grease-free cloth. To make good preparations, slides must be free from grease. Great care must be taken when polishing cover-slips, because they are very fragile and very expensive – so expensive, in fact, that the amateur will probably wash them for re-use.

Wet mounting

The easiest and simplest preparation to learn to make is the wet cover-slip preparation. This is used to examine any fluid specimen, such as pond water or blood. The technique is very simple, although a little practice is required. Place a small drop of the fluid to be examined in the middle of a clean slide. For a $\frac{7}{8}$-inch square cover-slip, the drop should be about as big as a split-pea. Lower a cover-slip gently on to the slide, either with the fingers or with the aid of a dissecting needle. Allow the fluid to spread out between the two pieces of glass without applying pressure. The specimen is then ready for examination. Simple wet preparations cannot be kept for very long because they dry up. To prevent this, the edges of the cover-slip must be sealed. A mount of a liquid preparation sealed in this way is not completely permanent, but it can be kept for quite a long time. The easiest method of sealing the edges is with petroleum jelly. For successful sealing, avoid having too much fluid under the cover-slip, otherwise the fluid oozes out at the edges. A more permanent mount can be prepared with the mounting medium *Canada balsam,* although to do this neatly requires practice. The balsam takes some days to harden completely.

Dry mounting

When mounting dry specimens, such as insect wings, foraminifera, and powders, use *gum arabic*. Spread the gum lightly on the slide and when tacky, place or sprinkle the specimen on the surface and allow the gum to harden. When hard, drop a little Canada balsam on the specimen and lower a cover-slip on to it. This will only work if the specimen is very thin.

A cell must be used for thicker dry specimens to separate the cover-slip from the slide. Aluminium cells are suitable and can be obtained in various thicknesses. They are cemented

Ringing to make a permanent preparation

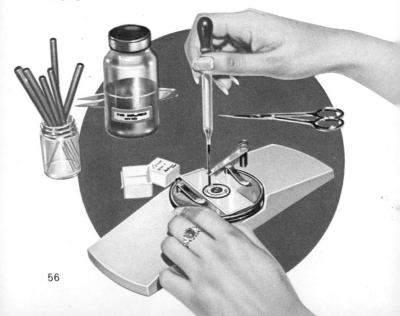

to the slide with Canada balsam. Spread gum in the cell centre and when tacky drop the specimen on to it. Spread a ring of balsam round the rim of the cell and lower the cover-slip into position. In this way, permanent dry mounts can be made.

Mounting media

There are many types of mounting media, which are divided into two main groups – the water-soluble media and the xylol-soluble media. Water-soluble media are used when mounting specimens directly from an aqueous solution such as pond water, but the mounts prepared in this way are not permanent unless the edges of the cover-slip are sealed with a suitable cement. The two most important water-soluble media are glycerin jelly and Farrant's medium.

When using xylol-soluble mounting media, such as Canada balsam or the synthetic mounting medium known as D.P.X., the specimen must be completely dehydrated and cleared (impregnated) with xylol. Preparations correctly mounted in these media are quite permanent.

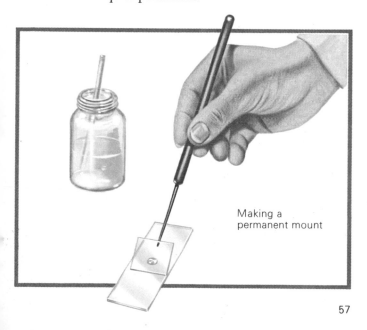

Making a
permanent mount

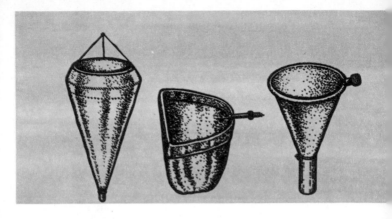

Nets for capturing specimens such as (*left*) plankton, (*centre*) pond life, (*right*) tiny insects

Examining pond life

The microscopy of pond life has a fascination for both new and experienced microscopists. Specimens can easily be collected from any pond or ditch which has held water for some time. A large number of minute organisms found in the pond are associated with weed. Therefore, all we need do to have an almost permanent source of material is to collect some bits of weed and to place them in a jar, together with some water.

A large jar filled with pond water, with some mud at the bottom and containing a selection of weed, can be kept at home as a permanent culture jar. Whenever material from other interesting sources is found, it can be poured into the jar to keep the culture going. Some of the organisms, such as the water fleas, or daphnia, will be readily visible to the naked eye, but many others are revealed only by microscopic examination of a wet preparation.

Plant life

Both freshwater and seawater weeds are called *algae*. Fresh-water weeds may be divided into two types – those with the cells either separate (for example, *Chlamydomonas* and *Euglena*) or in colonies (for example, *Pandorina* and *Volvox*);

and those in which the cells are joined to form long filaments (for example, *Conferva* and *Spirogyra*). The last type is the easiest to study. When a few strands of blanket weed, for example, are wet-mounted, the cellular structure can easily be examined both at low and high magnifications. But the smallest algae are less than ten microns in diameter and are difficult to examine. It requires special techniques rather beyond the scope of the amateur. Also included among the algae are the single-celled diatoms and desmids. Desmids have a greenish tint and are more delicate than diatoms. Diatoms are such ideal objects for microscopy that they are used as test objects for lens resolution.

A culture of pond water can be kept in a large jar.

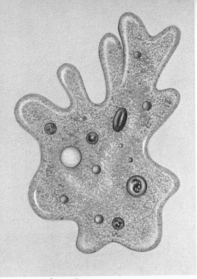

Amoeba

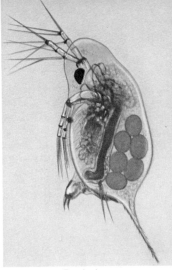

Daphnia

Animal life

Animal forms abound in most ponds and the simplest are the single-celled animals, which are called protozoans. The simplest one of all, is the amoeba, which is constantly changing shape and throwing out 'false feet', or pseudopodia. A close relative of the amoeba, *Actinophrys sol*, has rigid pseudopodia radiating out from the centre giving it a characteristic sun-like appearance. *Actinosphaerium* lives amongst the surface weed and in winter changes itself into a cystic form – that is, with a hard outer coat.

Of the ciliated protozoans, *Stentor* is common and often trumpet-shaped, with a mouth ringed with constantly moving cilia. *Vorticella,* another interesting ciliate, can be found attached to submerged plants. *Paramecium* may also be found, oval in shape and covered with cilia, which serve both to collect food and to propel the organism through the water. Many interesting forms of life are to be found in the sludge at the very bottom of the pond. To examine the sludge, pipette a small quantity on to a slide and place a cover-slip carefully on top. We will probably find small round-worms called nematodes, together with flat-worms up to a centimetre in length. We will also find other worm-like creatures with a discernible head, neck, and body. They are the Gastrotricha.

Only the female of the Gastrotricha can be found, usually with its posterior half-filled with eggs.

To describe all the minute organisms to be found in a pond would more than fill a book of this size, and plenty of more detailed books are available. The Rotifera, for example, can provide many months, or even years of pleasurable study. These animals vary greatly in shape, colour and size – from about 50 to 200 microns in diameter. They can be found swimming or crawling, or even glued by a sticky secretion to weeds and sticks. Other organisms to look for are the commonly found Crustacea, which are divided into three main groups – the Cladocera, Ostracoda and Copepoda. The Cladocera include the Daphnia, and the Copepoda include the *Cyclops* species.

Among the smallest forms of animal life in the pond are the Foraminifera. They are single-celled organisms which are surrounded by a coat of a chalk-like substance. Indeed, most of the chalk around our shores is formed from foraminifers deposited from the sea, where they are found in enormous numbers.

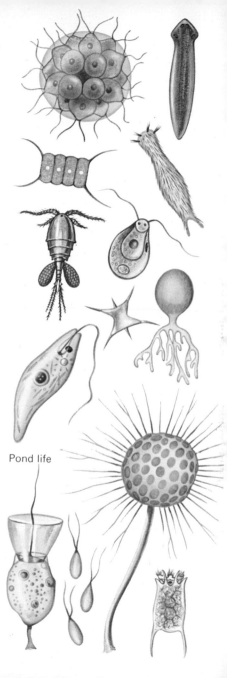

Pond life

Examining plants

When we come to examine a botanical specimen, we are often presented with the problem that the specimen we wish to examine is too large to get under the microscope. We must therefore revert to examining a small portion of the whole object to find out more about its composition. To do this, we must make the sample piece not only small enough to get under the microscope, but also thin enough to allow light to pass through it. If we take even a thin leaf and put it under the microscope, we can see little of its internal structure. We must, therefore, cut and examine a thin section of it.

Cutting sections

For the simplest sectioning technique, we need a piece of pith from the centre of a stem and a razor blade – preferably, for safety's sake, single-edged. First make a longitudinal slit in the pith and insert the leaf edgeways into the slit. Hold the pith firmly in one hand and, with a very sharp blade, cut the thinnest possible transverse sections of the supported leaf.

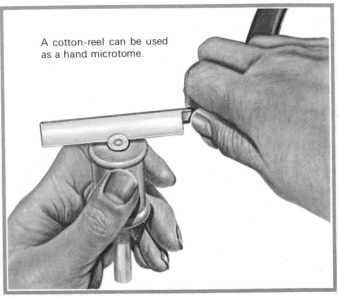

A cotton-reel can be used as a hand microtome.

A hand microtome

Supporting the leaf in this way makes it easier to get a clean-cut surface. Mount these slices on a slide in a water-soluble medium and examine them directly with the microscope. We can see that plants, like animals, are composed of millions of tiny cells, each with its own cytoplasm and nucleus.

The amateur requires very little apparatus to obtain quite adequate sections of botanical material. Even such a simple object as a cotton reel can become a useful section-cutting device, or *microtome*. Push a piece of plant stem up the centre hole and shave off the protruding part. Then push the stem gradually through the hole and shave off thin sections using the end of the reel as a guide for the razor blade.

Hand microtomes are rather similar to a cotton-reel with a screw thread up the centre and a screwed pushing device. Each turn of the screw-thread advances the specimen by a set amount. Additional support is provided by embedding the specimen in a suitable medium so that it can be easily sectioned. A technique for doing this with biological material is described later (see page 98).

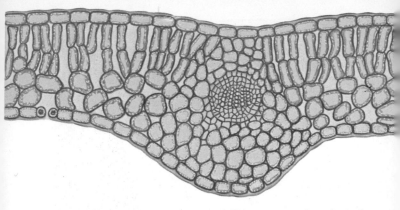

Transverse section of a leaf

Pollen grains (*left*)

Investigating cell structures

When we examine the sections under the microscope, we can easily distinguish individual cells. Each cell is bounded by a thin, colourless wall called the cell *membrane*, which surrounds the *cytoplasm*. The *nucleus* of the cell can be seen within the cytoplasm. Spread throughout the cell is the *cell-sap,* either dispersed or in small droplets. The cell-sap is the coloured part of coloured plants such as the geranium or beetroot. Closer examination of the cells of plants may reveal within the cytoplasm small green specks of *chlorophyll*, which is found in the leaves and stems of all

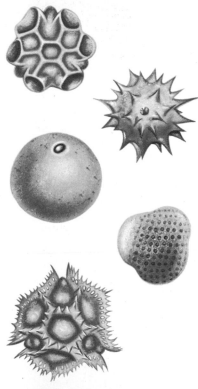

healthy plants. *Starch* granules and *oil* droplets can also be found in the cytoplasm of some plant cells.

If we examine a leaf bud from, say, a horse-chestnut tree, we find that it contains a large number of very small cells, all complete within the bud. And if we examine an older leaf, we find that it is composed of many large cells. We have therefore discovered that the leaf grows by the enlargement of its cells and not by the production of new ones. But if we examine cotton filaments, for example, we find that the reverse is true. The filaments grow because, when each cell reaches a certain stage, the nucleus divides and two new cells are formed. These two cells grow and divide further, and so on. With the microscope we can observe both types of growth – cellular enlargement and cell division.

Botanists are not confined to looking at sectioned material. Much can be done without such preparation. The examination of pollen grains, for example, can reveal a wealth of information. Even simply pulling off the skin of a leaf or plant and mounting it, is a useful technique.

Transverse section of root of a young plant

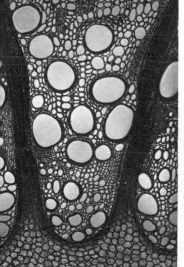

Cross section of a plant stem

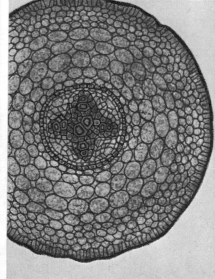

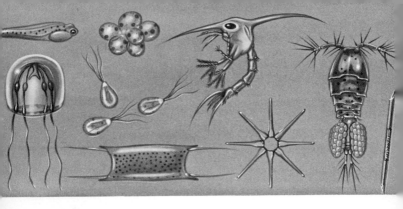

Plankton

Examining marine life

The methods used to examine marine life are naturally almost identical to those used to examine pond water, and in many cases the organisms found in the sea are closely related to those found in fresh water. One of the major attractions of sea water for the microscopist is in the examination of plankton, even if gathering the specimens requires the use of a boat and a plankton net. Plankton consists of millions of minute organisms which occupy the upper layers of the water. To examine specimens under the microscope, we must first discard the larger creatures such as baby shrimps, worms and fish and make wet preparations from the remainder.

Examination of the plankton reveals many diatoms and foraminifera upon which many of the larger organisms feed.

Foraminifera

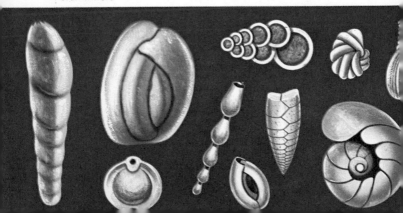

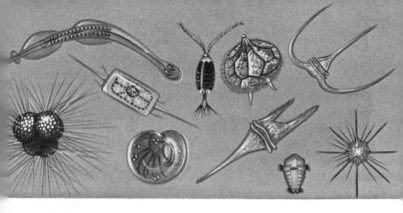

In summer, a jelly-like flagellate called *Phaeocystis* will abound. This is the organism that gives off an unpleasant smell and clogs the nets. Other flagellates are the luminous and transparent *Noctiluca,* inside which it is often possible to find diatoms, and the bell-shaped dinoflagellates such as *Tintinnopsis.* Small jellyfish and many other creatures will have to be preserved, by adding up to 10% formalin to the water. They can then be removed with a pipette, dropped on to a slide, and mounted in a water-soluble medium such as glycerin jelly.

However, we need not confine ourselves to examining small organisms. We can prepare sections of seaweed, and mount fish scales and fins. As with pond life, there is no limit to the number of different specimens that can be obtained. A few gallons of seawater can provide years of study and enjoyment.

House fly

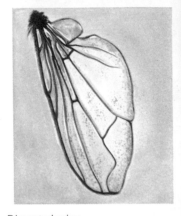

Dissected wing

Examining insects

Entomology – the study of insects – is another subject that relies a great deal on the microscope for its existence. It is often necessary to resort to the microscope to establish the identity of a particular insect species.

Catching specimens

From the point of view of capture, insects must be divided into two groups – those with and those without wings. Winged insects are usually caught with a net or by spraying with an aerosol insect-killer. Spraying is a quick and effective method of catching specimens, especially indoors. Not only does it catch the insect but it also kills them without damaging their appearance. Non-winged insects such as fleas, lice, mites and ticks are usually collected directly from the animal host or, in some cases, by dissection of the host's nest. Dead insect specimens can be stored in test-tubes with cotton-wool plugs.

Preparing and mounting insects

In the case of small insects, such as sand-flies or midges, direct preparations can be made of the whole insect on a micro-slide, provided that plenty of mounting medium is used. Some insects, such as fleas, have a hard outer covering

Mite

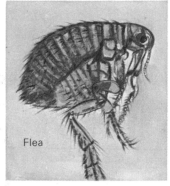

Flea

Simple
dissecting
microscope

of a substance called *chitin* which must be softened before
they can be mounted. This is done by dropping the specimens
into hot 10% potassium hydroxide (caustic potash) for ten
minutes, and then washing them in water for a few minutes.
The specimens must then be soaked thoroughly in spirit to
remove all the water before they are cleared in xylol and
mounted on a slide in Canada balsam.

Larger insects need to be dissected before proper microscopical examination can be carried out. For this purpose, a
simple single-lens dissecting microscope is of great value
because magnifications of about ×5 to ×10 are ideal for

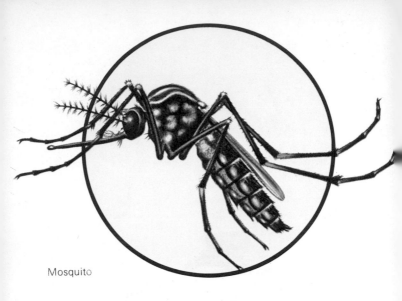

Mosquito

dissection. With practice, the head, legs, wings, and so on, of an insect such as the mosquito can be removed and mounted. Chloral hydrate is a most useful mounting medium because insects and their dissected parts can be directly mounted in it. The slides must be ringed to make permanent preparations.

Larger insects still, such as house-flies, are often best kept whole as pinned mounts and examined with the dissecting microscope. To mount one, take a small entomological pin and push it through the middle of a piece of card measuring about 2 cm. × 1 cm. Then push the point of the pin into, but not right through, the thorax of the insect. Push a second, larger pin through the card in the opposite direction to the first pin, and into a cork. The specimen is then permanently mounted for examination.

Preparing and mounting larvae

An insect such as a mosquito lays its eggs in water, and both larvae and pupae can be found there. The larvae of black flies, or buffalo gnats, can be found attached to rocks or logs in fast-running streams. Most larvae and pupae require some form of preparation and, as an example, the steps in preparing an ordinary house-fly larva for examination will be given.

The larva of a house-fly is whitish in colour and about 1·25 cm. long. It is pointed at one end and flat at the other. On the flat end are two small brown areas, which are the spiracles, or breathing tubes. The first operation is to kill the maggot and soften the outer skin by dropping it into hot 10% caustic potash, as in the case of the flea. Death is instantaneous, and the immediate effect is that the larva straightens. After five to ten minutes, remove it from the caustic solution with forceps, rinse in water, and place it on a microscope slide.

Hold the body of the larva gently with forceps, cut off the flat posterior end with a sharp razor as near to the end as possible. For this operation, a small dissecting microscope is useful, although it is not essential. Transfer the thin, round disc containing the two spiracles to a 70% spirit solution to prevent it drying up while the rest is being prepared. The next step is to remove the body contents. Gently hold the pointed anterior end and, with the side of a dissecting needle, squeeze out the contents until only the translucent skin remains. Take great care not to pull the skin apart. If this happens, start again with a fresh specimen.

Place the cleaned larval skin in the 70% spirit solution with the end plaque for a few minutes. Then transfer them both to absolute alcohol and give them several changes over a period of ten to fifteen minutes to ensure complete dehydration. When dehydration is complete, clear the specimen in xylol or, better still, beechwood creosote, and mount it under a cover-slip in Canada balsam, with the end plaque alongside the main body, spiracles uppermost.

Microscopical examination of the larva reveals a wealth of detail, such as the mouth hooks at the anterior end and the spiracular slits within the spiracles.

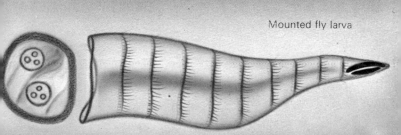

Mounted fly larva

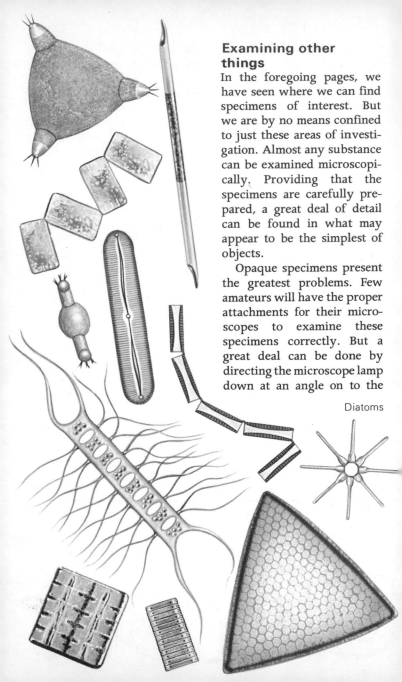

Examining other things

In the foregoing pages, we have seen where we can find specimens of interest. But we are by no means confined to just these areas of investigation. Almost any substance can be examined microscopically. Providing that the specimens are carefully prepared, a great deal of detail can be found in what may appear to be the simplest of objects.

Opaque specimens present the greatest problems. Few amateurs will have the proper attachments for their microscopes to examine these specimens correctly. But a great deal can be done by directing the microscope lamp down at an angle on to the

Diatoms

microscope stage. By this technique, newspaper photographs, for example, can be made to reveal their structure of thousands of tiny dots. Cheese, hairs, fibres, shells, and even transistors and tiny integrated electronic circuits make fascinating viewing.

The amateur microscopist often becomes disheartened not because of lack of material, but because of bad specimen preparation. The examination of some specimens – blood, for example – which seems exciting at first, turns out to be quite dull and very disappointing if the specimens are not properly treated before they are examined. As will be seen later, some of this treatment is quite complex, but much of it is quite within the scope of the amateur.

Desmids

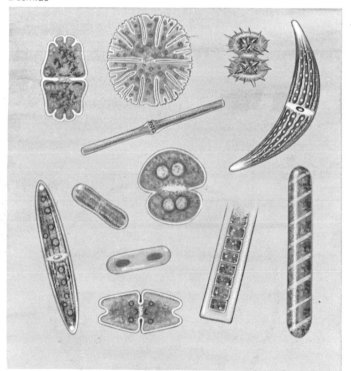

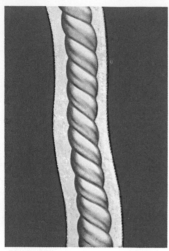

Fibre of cotton

A newspaper photograph can be seen to be composed of tiny dots.

Specializing

The microscopist has so many specimens and techniques at his disposal that he cannot possibly be an expert at them all. Even the amateur microscopist should be prepared to specialize in one particular branch of the science. The branch he chooses will often be dictated by the local environment – the marine biologist, for example, is obviously in a far better position for collecting specimens if he lives within easy reach of the sea.

Initially, some knowledge of each subject can be gained from simple preparations. But as progress is made, more specialized techniques will be required, each one related in some particular way to its own branch of microscopy.

Hair of kangaroo

Hair of rat

Hair of bat

Drawer cabinet for slides

Lens cleaning with blower brush

Hardwood box for slides

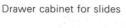

Looking after the microscope

A good microscope will last a lifetime if it is properly looked after. Dust causes wear of the mechanical components, and if left to settle on the lenses, it not only impairs their performance, but causes damage by scratching when they are cleaned.

Always keep a dust-cover in place when the microscope is not in use, and never leave an eyepiece out of the tube or dust will quickly settle on the objective lens at the bottom. Dust particles on the eyepiece will not materially effect the performance of the microscope. But dirt on the objective lenses, particularly greasy dirt such as a fingerprint or dried immersion oil, will have a considerable effect on the sharpness of the image.

Dust is best removed with a soft, grease-free brush or a brush with a small air-puffer attached (designed specially for lens cleaning), which can be obtained from most photographic shops. Greasy deposits, immersion oil or mounting medium must be removed at once with a little solvent such as ether or xylol. *Never* use alcohol because this dissolves away the cement used in the lenses. Final cleaning is best carried out with a little distilled water.

Storing slides

This soon becomes a problem to the microscopist, and several types of slide storage boxes are available. It is an advantage to choose a box into which the specimens can be placed immediately they are mounted, without the excess mounting medium around the cover-slip becoming stuck to the box. Boxes containing stacks of cardboard trays are quite good in this respect, especially if small cover-slips are used in the middle of the slide. The more robust wooden or metal boxes that hold slides vertically are ideal but only after the mounting medium is thoroughly dry.

Ordinary 5 × 3 inch card filing boxes are excellent for holding large numbers of slides. Obtain pieces of card about 5 × 4 inch and fold up the bottom inch. Stick down the edges of this folded portion with tape. It will be found that these cards will hold four standard 3 × 1 inch slides separated if required by staples through the folded part. In this way, hundreds of slides can be filed in one box.

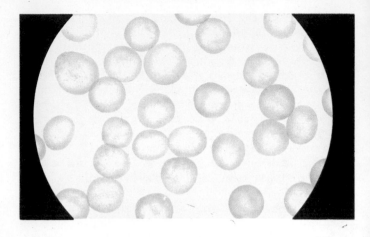

Unstained smear of blood

STAINING TECHNIQUES

Why we need to stain

Having described the simple techniques, we can progress to those that are more complicated and probably beyond the scope of most home microscopists. Even so, many of these techniques are not too difficult and can be modified to suit particular requirements.

Simple specimen preparations will reveal only a certain amount of the information we may need. As we have already seen, it is necessary to cut sections of some materials to see anything of them at all. Many specimens, particularly in biological work, are completely transparent, and it is difficult to see the structures we know to be there. If we place a drop of blood on a micro-slide, put a cover-slip on top, and examine the preparation under the microscope, we shall see countless tiny cells tumbling one over the other, but even the highest magnification will reveal little additional information. We have reached the limit of the simple preparation.

Making smears

The next step is to arrange the cells in such a manner that we can see each one individually. For this we need two clean, grease-free slides and a small drop of blood. Place the drop of blood at one end of one slide, and place the other slide at an angle of about 60° to the first one. Gently pull the end of the second slide along the face of the first until it touches the drop of blood, which then flows along the edge. Then push it forwards. As it moves forwards, it pulls the blood with it. We now have a film of blood on the first slide which should be just one cell thick. If we examine this blood smear under the microscope, we find that it is almost impossible to see the cells. Only a faint outline is discernible. To see the structure of the cells, they will have to be stained with

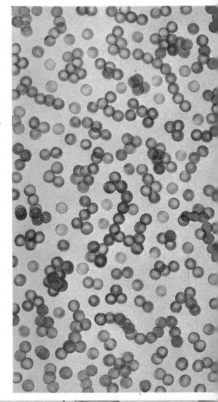

Blood in a wet preparation (*right*)
Stained smear of blood (*below*)

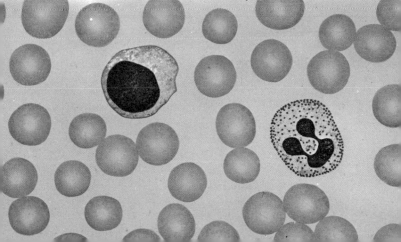

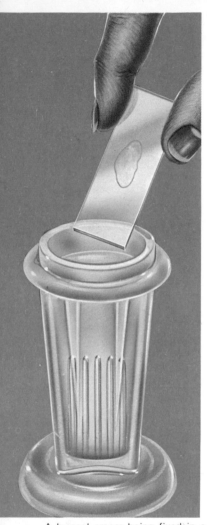

A buccal smear being fixed in a Coplin jar. The moulded ridges will hold several slides upright.

various dyes, which make the cells appear coloured when we examine them. By using different dyes, usually in combination, we can stain different cell elements in different colours, and in this way build up a colour picture of their construction.

Fixing smears

Before we can stain the cells, we must treat them with a form of preservative called a *fixative*. Without this, they would wash off the slide or be ruined by the water in the stain. All cells contain certain salts. If the fluid surrounding the cells does not contain salts of a similar concentration, the cells will burst because of the pressure difference between inside and outside. These pressures are called *osmotic pressures,* and salt solutions of the correct concentration are called *isotonic*. Fixing must preserve the cells and protect them in subsequent processing.

There are many different fixatives, each used for a different process; some stains do not work properly if the wrong fixative is used. The fixatives used for smears are usually rather fierce and rapid, designed to precipitate the proteins in the cells.

Ordinary absolute alcohol, for example, can be used as a fixative. The usual fixative for blood films is absolute methyl alcohol. Cells such as epithelial cells, easily scraped from the inside of the cheek (a *buccal smear*), are best fixed in a mixture of equal parts of absolute ethyl alcohol and ether. In both cases, fixation is almost instantaneous. After about a minute, the smears can usually be removed from the fixative, allowed to dry, and stored in this condition. The fixatives used for bulk pieces of tissue are much more gentle in their action than those used for smears and take much longer to work.

The fixation technique used for a blood smear is slightly different. The stain to be used, called *Leishman stain,* is made up as a solution in absolute methyl alcohol. Pouring the stain on to the slide therefore fixes the cells. The stain does not work in this concentrated form. It only starts to stain after it has been diluted with water which has been corrected for alkalinity.

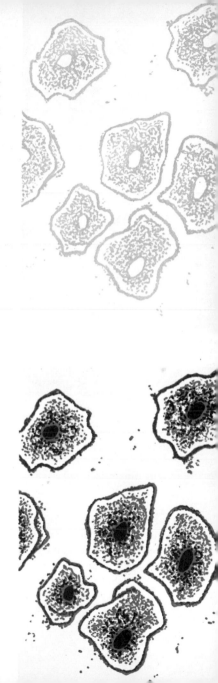

Buccal smear cells unstained (*above*)
Buccal smear cells stained in methylene blue

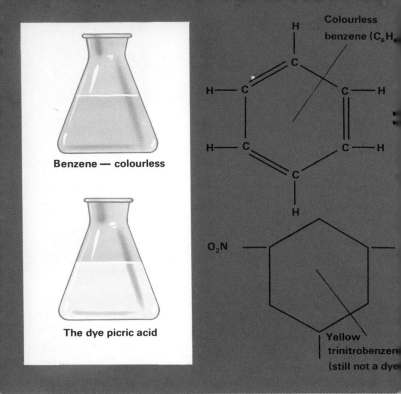

Benzene — colourless

The dye picric acid

Colourless benzene (C_6H

Yellow trinitrobenzen (still not a dye

Kinds of stains

Stains were first used in the preparation of specimens for the microscope in the 1700s. Leeuwenhoek used *saffron* to stain muscle fibres as early as 1714, and it is reported that Sir John Hill used *carmine* to stain specimens of timber in 1770, although a century passed before this was used again.

In 1856, the so-called *aniline* dyes used in the textile industry were introduced to biology, and a new era of specimen preparation and discovery began. The microscopical stains used today are modified commercial dyes, usually made to a greater degree of purity. In some stains, however, impurities are necessary for a reaction, probably acting as mordants (page 84). The famous stains made by Grübler in Germany before World War II were said to contain only the best impurities!

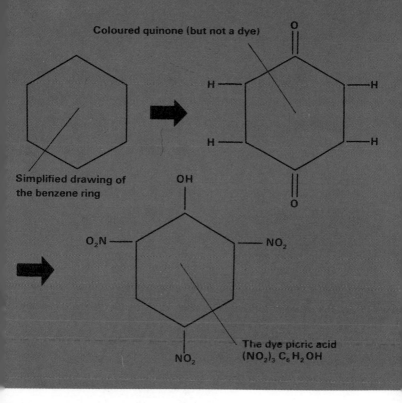

Adding molecules to the basic benzene ring produces coloured dyes.

Nearly all dyes used to be called aniline dyes in spite of having little or no connection with aniline. Almost all of them are derivatives of benzene. Most dyes are synthetic, although a few natural ones are still in use. *Haematoxylin* and *carmine* are the most important natural dyes for tissue staining. *Orcein* and *litmus* are other examples of natural stains. Carmine is produced from cochineal, which is extracted from the cochineal insect, *Coccus cactii*. Orcein and litmus are extracted from lichens.

Haematoxylin, the most widely-used nuclear stain in biology, is extracted from the wood of the logwood tree of Mexico, *Haematoxylon campechianum*. Haematoxylin powder itself has no ability to stain until it has been oxidized to haematin – a process called *ripening*. Haematoxylin solutions are oxidized either by exposing them to air, preferably direct sunlight, for periods of six to eight weeks or by adding an oxidizing agent such as mercuric oxide.

Principles of staining

The theory of staining is complex, but there are two main principles. When stain is more soluble in the tissue than its own solvent, it becomes transferred from the solvent to the tissue. Staining also depends on *adsorption* – the property of the body to attract minute particles from its surroundings.

The strict rules of chemistry do not fully apply to staining. Acid haematoxylin, for example, does not stain the basic elements of the cell but the acidic nucleus instead. Also, if staining were purely a chemical reaction, it should be possible to use stains to exhaustion. But this is very often not the case. No matter how much you dilute the stain or how long you use it, it still keeps functioning.

There are three staining techniques in general use: *non-vital* staining, which is the staining of dead tissues; *vital* staining, which involves the staining of live tissues; and *histochemistry*. In histochemistry, the most modern of the three techniques, colour reactions are used to demonstrate the various elements of the cell or tissue. These reactions are often triggered off by colourless solutions. The best-known histochemical reaction is the *Feulgen reaction,* in which colourless Schiff's reagent is used to stain the basic protein of life (DNA, or deoxyribonucleic acid) shades of reddish-purple.

Non-vital staining

Most microscopists will be concerned only with non-vital staining, usually of fixed tissues. Some stains, such as methylene blue, stain tissues directly, and a simple solution of the powder will have the ability to colour tissues. But many stains require the presence of a third agent, a *mordant,* which forms a link between the element to be stained and the stain. The stain will not link directly with the tissue but it will with the mordant. The mordant will link with the tissue and the stain so that the net result is stain linked to the tissue via the mordant. The compound formed between the stain

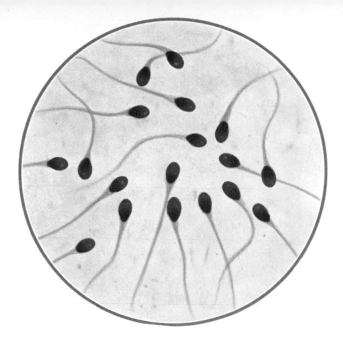

(*Above*) The Feulgen reaction has stained the sperm heads red.
(*Below*) Vitally stained reticulocytes in the blood

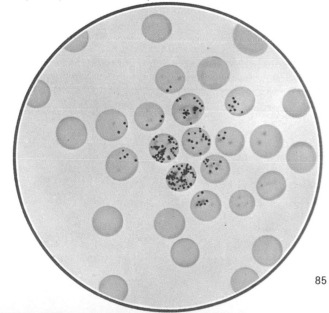

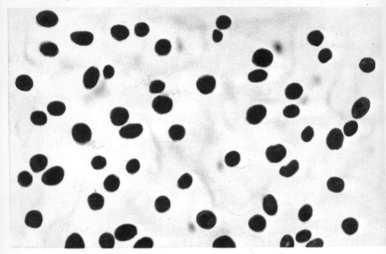

Cell nuclei stained with haematoxylin

and the mordant is called a *lake*. Many of these lakes are unstable, and for this reason some tissues to be stained must first be soaked in the mordant and then in the stain.

There are other substances used in staining reactions which are not true mordants but are merely accelerators or accentuators. They help the reaction but do not take part in any chemical union between stain and tissue. An example of a substance which does not act as a mordant but helps in the staining reaction is phenol in the carbol fuchsin solution used for staining the organisms which cause tuberculosis. The phenol probably alters the bacteria in some way to allow the red dye to penetrate.

What are stains?

All normal stains are salts composed of an acid and a base. They are termed *acidic, basic* or *neutral,* according to which part is coloured. For example, acid dyes have the acid component coloured and the base colourless. Most acid dyes stain the basic cytoplasm of cells, whereas most basic dyes stain acid components such as nuclei. To stain a whole cell both acid and basic stains must be used, in some cases

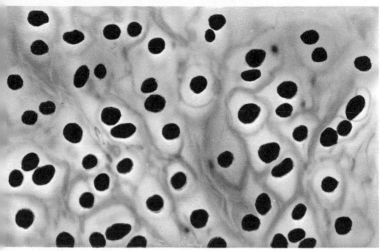
Cell nuclei counterstained with eosin

together and in other cases one after the other.

Neutral dyes, on the other hand, have both basic and acidic components coloured. They are prepared from the precipitate which forms when basic and acidic dyes are mixed in aqueous solution. This precipitate is normally soluble only in alcohols. The best examples of neutral stains are the *Romanowsky* group (that is, *Leishman* and *Giemsa* stains), which are formed by the interaction of eosin and methylene blue. These stain both acid and basic components of the cell. However, some components (called *neutrophilic*) take up both dyes simultaneously.

Polychromasia

Polychromasia is another phenomenon of staining. Although most dyes stain tissues varying shades of their own colour, certain dyes (called *metachromatic*) can stain different cell components different colours. For example, toluidine blue will, under certain circumstances, stain mucin red, while the rest of the tissue is stained blue.

We can see from the above that staining is a very complex subject. Even so, only a relatively few stains are in regular use. The commonest of all is probably the haematoxylin and eosin stain in which the cell nuclei are stained bluish-purple (haematoxylin) and the cytoplasm various shades of pink (eosin). As we shall see later (page 104), simple staining is not beyond the scope of the amateur. For simple staining methylene blue in aqueous solution is extremely useful.

Stained specimens

Staining techniques open up a whole new world of microscopy. They are widely used by both the zoologist and botanist. The rather indistinct piece of onion skin, for example, in which the cells are almost invisible, looks very different when stained with haematoxylin, because the nuclei and cell membranes become clearly visible. Amoebae found in the pond, once fixed and stained, reveal many structures which are quite invisible in the fresh specimen. In the diagnosis of diseased tissues, staining techniques are of the utmost importance.

Staining a buccal smear

There are illustrations on page 81 showing epithelial cells of a buccal smear, some stained and some unstained. The staining method used is very simple.

(*Opposite above*) Amoeba stained with iron haematoxylin
(*Centre*) Onion skin stained with haematoxylin
(*Below*) Liver stained with haematoxylin and eosin

Bone marrow stained with Giemsa stain

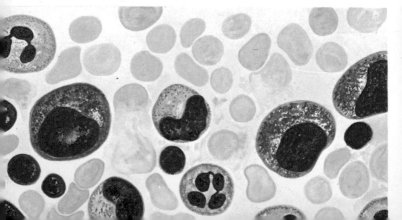

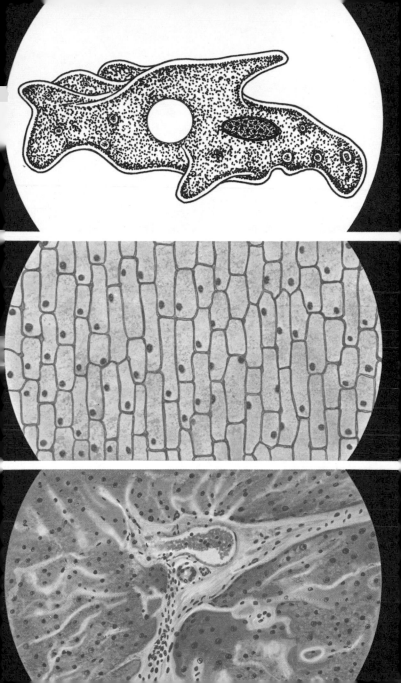

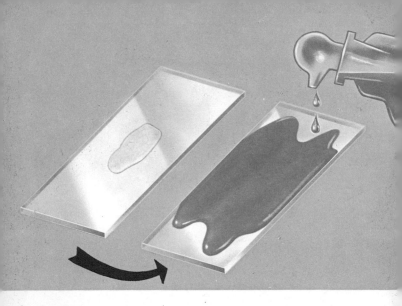

The smear is prepared by scraping cells from the inside of the cheek with an instrument such as a spatula. Very little pressure is required because they scrape off quite easily. Spread out the debris on the spatula on to a slide so as to cover an area of about one to two square centimetres. Drop the slide immediately into a jar containing a mixture of equal parts of ether and alcohol (the fixative). An ideal jar for the purpose is a Coplin jar, which will hold a number of slides upright between the ridges moulded into its sides.

After a few minutes remove the slide from the fixative, rinse it first in alcohol and then in water, and place it on a staining rack, which consists of two pieces of rod placed about two inches apart across a sink. Pour 1% methylene blue in aqueous solution on to the slide and leave it for five minutes. Tip off the stain and rinse the slide in running tap-water to remove the excess stain. Then prop the slide up on end and leave to dry. Once dry, it is ready for examination, but only with the low-power or oil-immersion lenses.

For satisfactory results with the high, dry lens, a cover-slip must be used. Mounting a cover-slip on to the slide with Canada balsam is very simple, although some practice is required if air bubbles are to be avoided. Put a drop of the

Steps in a simple staining procedure: staining with methylene blue, rinsing in water and drying.

mounting medium on to the smear and gently lower a clean cover-slip on to it. Do not press the cover-slip down but allow its own weight to spread the balsam. If the balsam is too thick, thin it down with a little xylol.

Microscopical examination of the slide will show many large, flat, plaque-like cells, some with small and some with larger nuclei. Some elements of the cytoplasm can be clearly seen – the darker dots probably represent mitochondria. The nucleus appears mottled because the proteins it contains are precipitated during fixation. This is a very simple staining technique, but a comparison of the stained and unstained smears will clearly show its value.

Staining an onion skin

Let us now look at a slightly more complicated staining method – the staining of a piece of onion skin with iron haematoxylin. For this technique, a mordant is required (iron alum) followed by a simple solution of haematoxylin. The tissue is stained, unfixed, before it is mounted on the slide. The solutions required are 5% iron alum and 1% haematoxylin. The haematoxylin solution is prepared by grinding up 1 gram of haematoxylin powder in 10 millilitres of alcohol and adding water to make up the volume to 100 millilitres. These solutions can be obtained prepared ready for use.

Peel a piece of skin about a centimetre square from an onion and place it in a small dish. Pour on the iron alum solution and leave it overnight. Next morning pick up the piece of skin by holding it gently at one corner with a pair of forceps and transfer it to a dish of water. Gently rinse it by waving it backwards and forwards in the water for a few seconds. Then transfer it to a third dish containing the haematoxylin solution. Take care not to carry over any of the iron alum solution, otherwise the haematoxylin will immediately turn black and be ruined. Leave the specimen in the haematoxylin solution for the same length of time that it was in the iron alum. By then it should be quite black in colour.

This treatment will almost certainly have overstained the specimen, and it will be necessary to remove the excess stain – a process called *differentiation*. This is done by alternately rinsing the specimen in the iron alum solution and water until the desired colour is reached. The colour desired will depend on which elements need to be stained and on personal preference. But generally it will be best to leave the specimen in the differentiator until it is dark grey.

To mount the stained specimen, we have a choice of two methods. The easiest one is to tease out the tissue flat on to a micro-slide with dissecting needles and drop on to it a small amount of glycerin jelly or Farrant's medium, followed by a cover-slip. For a permanent preparation, the specimen should be mounted in Canada balsam. The tissue must first be dehydrated because Canada balsam will not mix with water. This is done by passing it in turn through 70% and 90% alcohol solutions and then absolute alcohol before transferring it to xylol. It can then be flattened out on a micro-slide and mounted in the balsam. This process requires some care and practice because the treatment with the alcohols and xylol will have made the specimen rather brittle. On no account allow it to dry up.

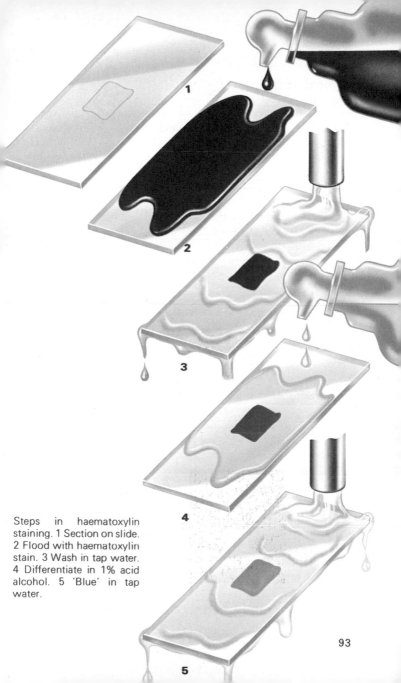

Steps in haematoxylin staining. 1 Section on slide. 2 Flood with haematoxylin stain. 3 Wash in tap water. 4 Differentiate in 1% acid alcohol. 5 'Blue' in tap water.

HISTOLOGY – THE STUDY OF ANIMAL TISSUE

Basic principles

We have already seen illustrations (page 88–9) of animal tissue which has been made into a microscope preparation and stained to show the cells. Such samples are prepared by taking sections of the tissue in much the same way as taking sections of botanical material. The sectioning and study of animal tissues is known as *histology*. It is such a vast subject that here we can only describe some of the fundamental techniques involved.

Before sections can be taken from soft tissues, the tissue must be *preserved* (fixed) and processed so that it is supported during the cutting process. We have all seen a butcher slicing easily through a piece of frozen liver. It cuts easily

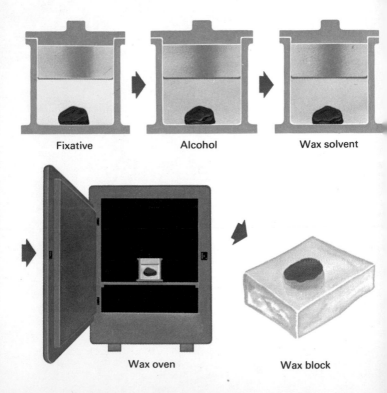

Fixative   Alcohol   Wax solvent

Wax oven   Wax block

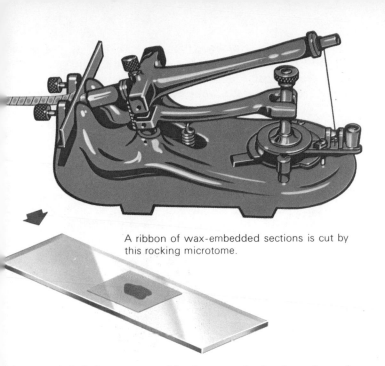

A ribbon of wax-embedded sections is cut by this rocking microtome.

because it is being supported by ice crystals that have formed within it. And one method of sectioning tissues for the light microscope is to freeze them before cutting. But by far the most important technique for supporting soft tissues for sectioning is that of embedding samples in paraffin wax.

Sections for satisfactory microscopical examination must be very thin. The average thickness is about 5 microns, although in general, the thinner the section the better. If photomicrographs are to be taken at high magnification, sections 3 to 4 microns thick are more suitable.

The pieces of tissue to be examined are fixed, dehydrated, cleared with a wax solvent, and finally impregnated with wax. The wax is allowed to set in a mould to form a block with the piece of tissue embedded in the middle of it. This block is trimmed and fixed to the chuck of a microtome which works on a similar principle to the bacon slicer. Thin slices are removed from the microtome with a camel-hair brush or forceps and floated on the surface of warm water.

The warmth helps to spread the wax evenly. The sections can then be picked up on glass slides and dried. We are left with sections of the tissue, still impregnated with wax, stuck firmly to the slide. The next step is to remove the wax and stain the tissue so that we finish up with the familiar microscope preparation.

Preparing the specimen

The processing of tissues or other substances follows the same general pattern whether it is a dead fly, a piece of plastic, or a diseased tissue removed by an operation in hospital. The fixation and dehydration stages can be omitted or modified for substances which will not decompose and

Pieces of tissue are selected and placed in fixative.

which do not contain water. But the techniques of embedding and sectioning are invariably similar. The embedding medium may be changed to suit the hardness of the material.

Fixing

Let us consider the simple specimen that we wish to section and examine microscopically. An ordinary house-fly will do, killed with an aerosol spray so as not to damage it. First we must decide what to fix it in. *Formalin* is the most widely used fixative in histology. It is the best general-purpose fixative, even though it does have certain disadvantages for some tissues and staining reactions. Formalin is a solution in water of the gas formaldehyde, which is soluble up to 40%. Concentrated formalin, therefore, contains 40% formaldehyde in water. We use as a fixative a 4% formaldehyde solution which is concentrated formalin made up as a 1 in 10 solution. It is quite satisfactory to make up this 4% formaldehyde solution in water.

Other common fixatives are *Bouin's fluid* and *Carnoy's fluid*. Bouin's fixative is a strong solution of picric acid in formalin. It is one of the best fixatives for the tissue element glycogen. Carnoy's fixative is a mixture of alcohol, chloroform and acetic acid. It is an excellent solution for preserving the nucleus and for fixing smears.

For our fly, we shall choose formalin. Simply drop the specimen into a bottle of the solution and leave it for at least twenty-four hours. The total time in the fixative depends on the size of the specimen and the fixative used. Formalin penetrates slowly but has the advantage that tissues left in it come to little harm. It can be used to preserve tissues indefinitely. Carnoy's fluid, on the other hand, is almost the opposite. Only small pieces of tissue should be placed in it, and they should be left for only a short period of time.

Dehydrating

After the specimen has been fixed, it must have all the water removed if it is to be successfully embedded in paraffin wax. This process of dehydration is the same as that used when sections are taken from water to xylol to enable them to be mounted in Canada balsam.

In this section mounting bath the water temperature is kept constant by a thermostatic control.

Remove the specimen from the formalin and place it first in a 70% solution of alcohol, then in a 90% solution, and finally in two changes of absolute alcohol. The length of time that the specimen stays in the alcohol solutions will depend on its size, but two hours in each solution should suffice for our fly. By this time, it should be fully impregnated with alcohol, and no water should remain.

Clearing and embedding

Transfer the specimen from the alcohol into the clearing agent (wax solvent), which can be one of a number of substances, including benzene, xylol and chloroform. Change the solvent twice over a period of time sufficient to ensure complete penetration.

From the clearing agent, transfer the specimen into the paraffin wax, which is kept in an oven a few degrees higher than the melting point of the wax. Change the wax three times to ensure that the clearing agent is completely removed from the tissues and molten wax substituted. When wax penetration is complete, embed the specimen in a mould containing molten wax and allow the wax to set. We can now break away the mould and have our specimen firmly supported in a solid block of wax, ready to be cut into sections with the microtome.

Sectioning

There are three major types of microtome, the *rocking* microtome, the *rotary* microtome and the *sledge* microtome. The simplest is the rocking microtome, which consists of an arm pivoted towards one end. A knife is held at right angles to the short end of the arm, to which the specimen is fixed. As the arm with the block of specimen attached is pulled up and down, a ratchet mechanism advances it towards the knife. As the block moves downwards, the knife slices thin sections from it.

With the rotary microtome, turning a large handle operates a mechanism which moves the block up and down against the knife and advances it by a predetermined distance each time.

The sledge microtome is the largest and most robust of the three and also the most versatile. It can cut both very small and very large pieces of tissue.

A rotary microtome

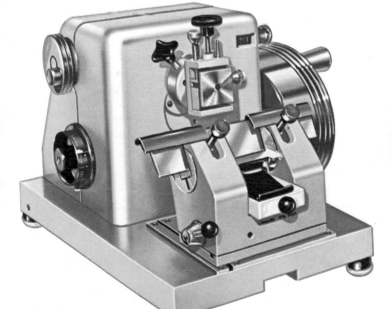

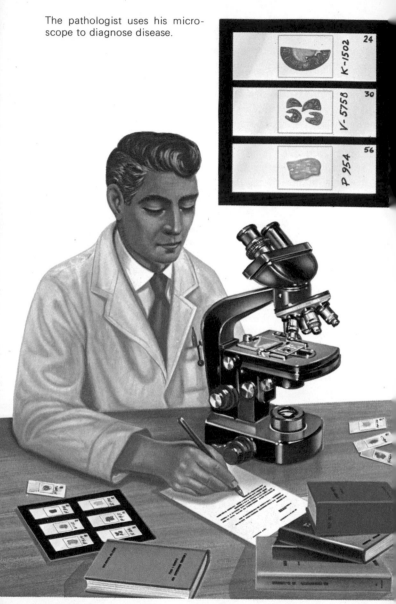

The pathologist uses his microscope to diagnose disease.

K-1502   24

V-5758   30

P 954   56

Preparing slides

Transfer the cut sections to a bowl of warm water, where they float and spread slightly. Any creases that may have appeared during the sectioning will disappear. Pick them up on micro-slides and allow them to dry, preferably in an oven set at about 37°C. When the slides are dry, the staining process can begin, commencing with the removal of the wax in xylol and proceeding through descending grades of alcohol to water. After staining and dehydration (back through the alcohols to xylol), the specimen can be mounted and labelled. The finished product is then ready for examination under the microscope.

Histology in disease diagnosis

Without doubt, the most important application of histology is in the hospital laboratory, where the pathologist diagnoses diseases by the examination of tissues removed by operation. Here the biopsy specimen is most important. A small piece of the suspected organ is removed and examined to confirm the presence of the disease before the more major operation is carried out. This process is particularly important in the diagnosis of diseases such as cancer.

The desire to obtain high-speed results in such cases has led to the development of different techniques in histology. As we have seen, the normal processing of specimens up to the finished slide may take several days – days which may make the difference between life or death to the patient.

In such urgent cases, the pathologist resorts to frozen sections. The piece of tissue is first fixed in hot formalin and then frozen to the chuck of a special microtome kept at a low temperature. The sections are cut in the same way that the butcher cuts his frozen liver. High-speed staining techniques have also been developed to the extent that it is now possible to get results in minutes rather than days. The life of the patient may well depend on the microscopical diagnosis of these sections.

In medical research, histological examination enables scientists not only to understand the ways in which tissues are attacked or destroyed, but also how they ward off bacterial invasions and are repaired.

BACTERIOLOGY – THE STUDY OF BACTERIA

The microscope also plays a major part in *bacteriology*, another division of science related to disease in Man. Bacteriology, or *microbiology* as it is sometimes called, is the study of the micro-organisms or germs which are to be found all around us. Bacteria can be divided into two groups: *pathogenic* – those causing disease – and *non-pathogenic* – those not causing disease. Many normally non-pathogenic organisms may cause disease under certain conditions. The bacteriologist uses the microscope to discover which of the many thousands of different organisms he is dealing with. Bacteria are only a few microns in diameter, and therefore their examination is confined to microscopes with high magnifying power. Bacteria are examined either live and unstained in suspension, or as smears dried on to slides and stained.

Examining live bacteria

The most convenient way of examining live bacteria in suspension is by the *hanging-drop* technique. Both the shape and the *motility* of the organism can be studied by this method. Motility is another way of distinguishing bacteria. Motile bacteria progress rapidly in a particular direction, whereas non-motile ones are moved only by the eddies and currents in the surrounding medium.

To examine a hanging-drop preparation, first make

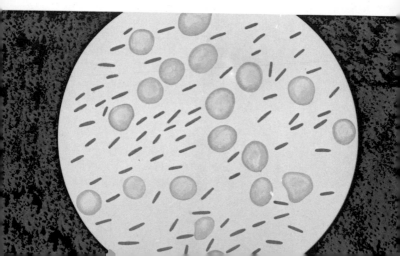

a well on a slide. Vaseline is an ideal substance to build up the walls of the well, which should be about 1 cm. across and 2 mm. deep. Then put a minute drop of the suspension on a cover-slip on the bench and invert the well slide over it. Gently lower the slide until the Vaseline ring touches the cover-slip encircling the droplet. Then turn over the slide until the cover-slip is uppermost. The drop should be hanging from the cover-slip in the centre of the well. It must not touch the slide.

To examine the specimen, first locate the edge of the drop with the low-power lens. (It will probably help to lower the condenser and shut down the iris; this is one occasion when it is best not to use Köhler

A hanging-drop preparation

Microscope view of the edge of the drop

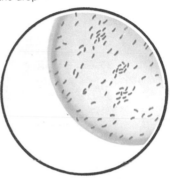

Bacilli (*left*)

Cocci (*below*)

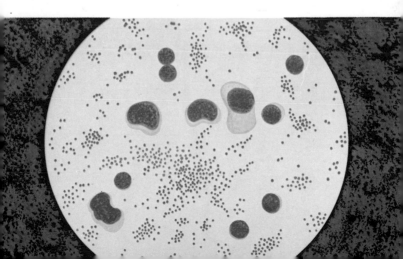

illumination.) With the high dry lens ($\times 40$), it should be possible to see the approximate size and general shape of the organisms. Motile organisms will be seen to swim to the edge of the drop and back again, but non-motile organisms will remain oscillating about one point. This technique can be applied to any fluid containing micro-organisms, including the pond-water culture described on page 58.

Examining bacterial smears
Staining
The most important stain in bacteriology is *Gram's stain*. This is used both to study the shape of the organisms and to classify them into one of two groups – *Gram-negative* organisms, which stain red, and *Gram-positive* organisms, which stain a bluish-black colour. The principle is that bacteria are stained with a dye such as methyl violet or gentian violet, fixed in iodine, and then treated with alcohol. This iodine fixation works only with the Gram-positive bacteria. The Gram-negative bacteria are decolourized by the alcohol and take up a red counter-stain such as neutral red.

Negative staining
Another interesting technique employed in the study of bacteria is *negative staining*. It has become more widespread by the use of the electron microscope in the study of viruses.

Gram-negative bacilli

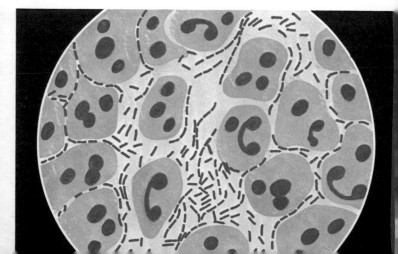

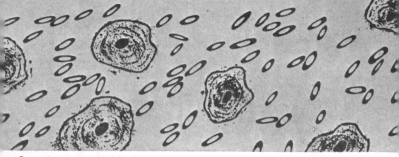

Capsules stained with nigrosin

The organisms to be examined are mixed with a substance which provides a dark, almost opaque, background when smeared on a slide. The organisms stand out as clear, unstained objects. Indian ink or the black dye nigrosin may be used for this purpose.

The technique is quite simple. Make a smear of the bacteria on a slide, and allow it to dry. Fix the smear on the slide by heating it until the slide is just too hot to bear on the back of the hand. Place a drop of 10% nigrosin on one end and, with another slide, make a smear of the nigrosin over the top of the bacterial smear (as for the blood film described on page 79). When the smear is dry, it is ready for examination, preferably under the oil-immersion lens. A permanent preparation can be made by mounting in Canada balsam. A modification of this method is to mix some of the bacterial

Gram-positive cocci

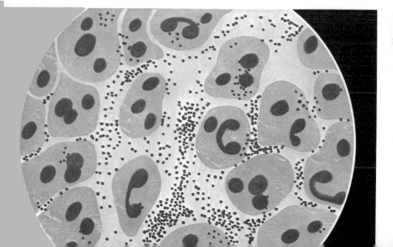

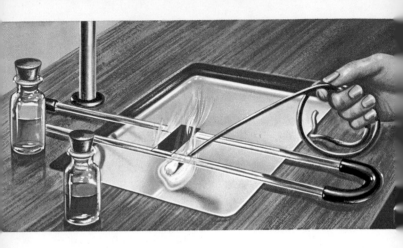

Heating the Ziehl-Neelson stain

suspension with a drop of Indian ink and make a smear of this. Allow it to dry, and examine as before. Either of these methods can easily be carried out by the amateur, provided that a suitable suspension of bacteria can be found.

Selective staining

In making a diagnosis, the bacteriologist may use some of the special selective staining techniques such as the *Ziehl-Neelson* method for the organism which causes tuberculosis and for related organisms, including those which cause leprosy. The final diagnosis then depends on finding the characteristically stained organisms with the microscope.

Tubercle bacilli, as they are called, do not stain with any of the usual dyes, probably because they have a fatty coat which prevents the stain from penetrating. To overcome this, a very strong solution of red basic fuchsin in phenol is used, and even this mixture has to be heated on the slide before the bacteria are stained. The advantage is that once the stain has penetrated the tubercle bacillus, it is almost impossible to decolourize it, even with strong alcohols and acid. Most other organisms, on the other hand, quickly become decolourized. Counter-staining with methylene blue completes the staining process.

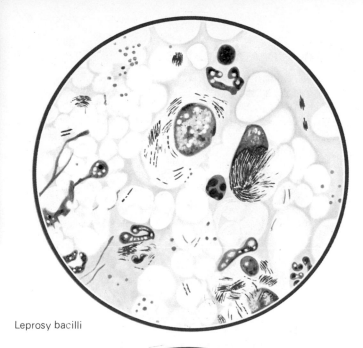

Leprosy bacilli

Tubercle bacilli

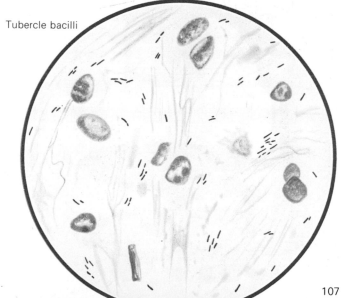

ADVANCED MICROSCOPY

The stereoscopic microscope

The *stereoscopic* microscope is designed to produce a three-dimensional image. It consists basically of two low-power microscopes set side by side and at a slight angle to each other so as to give a stereoscopic effect to the image. This dissecting microscope, unlike ordinary microscopes, makes

A zoom stereoscopic microscope

the specimen appear the right way round, so that any movements made under it are seen correctly.

Stereoscopic microscopes are restricted to quite low magnification, usually ×1·5 to ×25. A recent development is the zoom stereoscopic microscope, which gives variable magnification, but the degree of magnification change is usually limited. Many modern stereoscopic microscopes are designed for a specific purpose – for example, repetitive assembly work – and have only one fixed magnification.

The depth of focus of these microscopes is quite large because of the low magnification employed, and only a coarse focusing control is used. This is of considerable assistance in dissecting work because both hands can be used for the dissection, only occasional focusing being necessary.

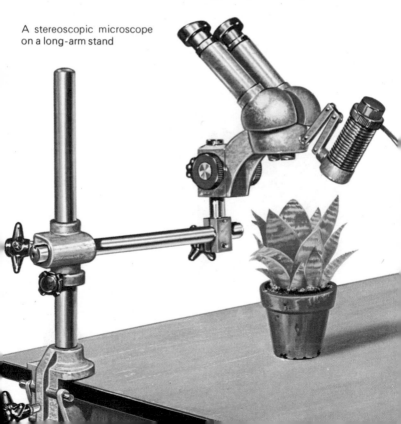

A stereoscopic microscope
on a long-arm stand

Dark-field microscopy

In *dark-field,* or *dark-ground*, microscopy, the specimen is illuminated by a special condenser that provides oblique light. The light will not pass up into the tube of the microscope unless it is deflected or scattered by the object. Only objects that have a different refractive index from the medium that surrounds them will do this. The objects show up as brightly-illuminated bodies against a black background.

The special condenser is either a paraboloid type or a concentric, spherical, reflecting type. It focuses a hollow cone of light at the plane of the specimen. The objective is positioned in the middle of the resulting hollow cone. In this way, no direct light can enter the objective lens from the condenser. Success with dark-field microscopy depends on an intense source of illumination and an accurately centred and focused condenser. The numerical aperture of the objective lens should not exceed 1·0 (see page 39). If it does, a funnel stop must be used in the lens to reduce the aperture.

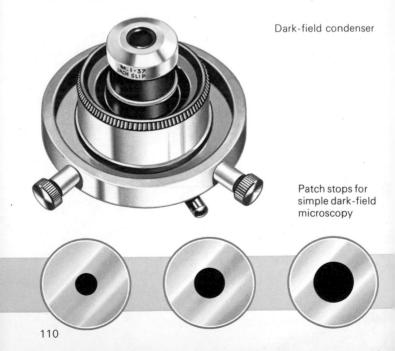

Dark-field condenser

Patch stops for simple dark-field microscopy

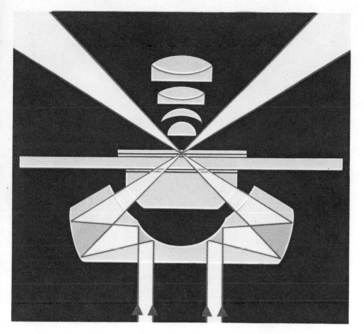

The path of light through a dark-field condenser

A simple method of producing dark-field effects with the low-power lenses is to use patch stops in an ordinary condenser. These can be made by sticking circular patches of black paper on to discs of glass and fitting them into the filter tray. The patch size (usually 10 to 12 mm.) should be such that the specimen is illuminated by a hollow cone of light.

This technique works with an ordinary Abbe condenser and a simple lamp, provided that the numerical aperture of the lens does not exceed 0·3 (that is, with a ×10 objective or less). With numerical apertures up to 0·65, a better condenser and a high-power lamp must be used, with a very accurately matched patch-stop. With patch-stop methods it is easy to flick from dark- to light-field illumination at will but results with the correct condenser are better than with patch-stopped condensers. For numerical apertures above 0·65, the correct dark-field condenser is essential.

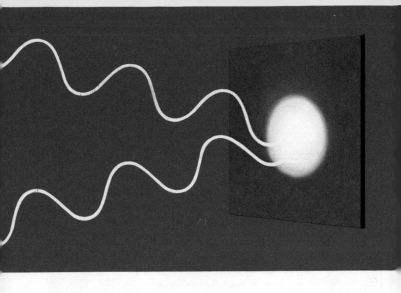

Two rays of light in phase produce light on the screen, but if one ray

The phase-contrast microscope

The greatest advance in biological microscopy in recent years has been brought about by the introduction of *phase-contrast* microscopy. Phase-contrast technique has made a major contribution to the study of living objects under the microscope. The problem with most biological specimens is that they are almost completely transparent, and staining to make them visible almost invariably kills them and alters their structure in some way. Phase-contrast enables contrast to be added to these normally invisible objects, making them quite visible down the microscope.

Theory of phase-contrast

As we have already seen, light travels in a sine-wave form, with characteristic amplitude, or wave height, and wavelength. The eye is sensitive to changes of wavelength (colour) and amplitude (brightness), but not to changes of phase when one light ray is retarded in relation to another. Biological objects are usually fairly uniform in thickness and have relatively uniform refractive index, but this refractive index is not necessarily the same as the surrounding medium. Light which passes through the specimen will be changed in

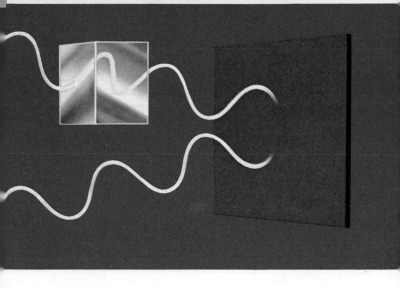

is out of phase no light appears on the screen.

phase compared with that which passes through the surrounding medium. But the specimen is not visible because these small phase changes are not readily detectable to the human eye.

Contrast in an unstained specimen depends on the ability of the specimen to alter the wave form by retardation or absorption. Light passing through the denser parts of the specimen will appear dark; also, light refracted by various degrees will be retarded and, because of destructive interference with rays not so retarded, appear darker. Thus contrast is brought about because light rays can *interfere* one with another.

If two light rays from an object strike a point on a screen, then the intensity of the resulting light will be equivalent to the sum of their amplitudes. On the other hand, if one of these rays has been retarded in some way by half a wavelength, then instead of getting increased light on the screen there will be no light. The bottom of one wave 'cancels' with the top of the other. Interference of light waves, therefore, may cause complete extinction of the light, or a reduction in intensity, depending on the relative amplitudes of the rays involved. Phase-contrast relies on enhancing the wave retardation of light that is refracted by the specimen. Light refracted by a specimen is usually retarded about a quarter of a wavelength. By increasing this retardation to half a wavelength, light passing directly through the specimen can be made to cause destructive interference with the refracted light, producing easily visible differences in amplitude.

How the microscope works

The basic requirements of the phase-contrast microscope are a phase-retardation plate in the back focal plane of the objective and an annulus, or ring, in or slightly below the substage condenser. The condenser annulus is designed to illuminate the specimen with a hollow cone of light. The objective lens forms an image of this light cone (that is, of the annulus) in its back focal plane, where the phase-retardation plate is situated.

The phase plate consists of a disc of glass with a channel ground into it which has been treated with material which absorbs light but does not retard it. This channel exactly matches the image of the condenser annulus. The remainder of the plate is treated with a thin film of a substance such as magnesium fluoride so that it retards light by a quarter of a wavelength. Light passing directly through the specimen and the channel is retarded a quarter of a wavelength less than the refracted light which passes through the remainder of the plate. Thus the light refracted by the specimen is now retarded a total of half a wavelength compared with the direct light passing through the channel. Interference of the deviated and undeviated rays produces the phase-contrast effects. The light-absorbing material reduces the brightness of the direct light and allows the contrast produced to be clearly seen.

The phase plates and annuli need not necessarily be rings, although the ring form is now generally accepted as the best configuration. Sometimes, for example, they are made in the form of a cross. The annulus match needs to be so exact that phase-contrast microscopes must be fitted with a centring and focusing condenser. In addition, it is also necessary to centre each individual condenser annulus accurately to its matching phase objective.

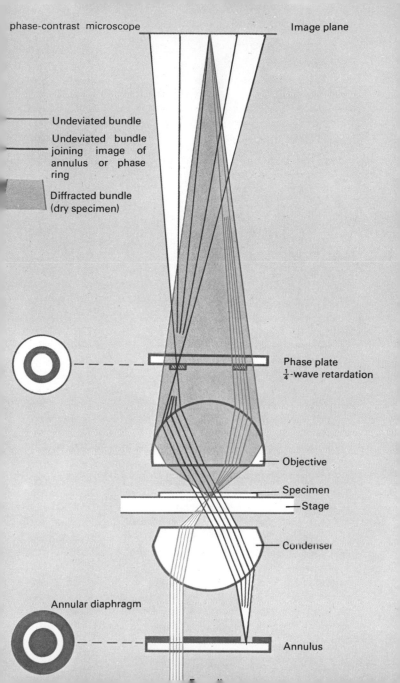

phase-contrast microscope

Image plane

Undeviated bundle

Undeviated bundle joining image of annulus or phase ring

Diffracted bundle (dry specimen)

Phase plate
$\frac{1}{4}$-wave retardation

Objective

Specimen

Stage

Condenser

Annular diaphragm

Annulus

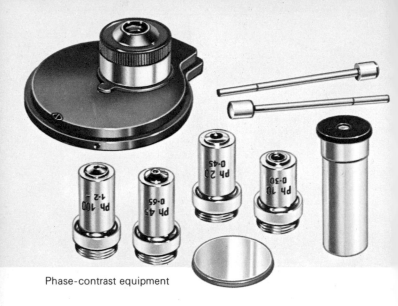

Phase-contrast equipment

Setting up the microscope

Setting-up the phase-contrast microscope is not too difficult if care is taken and the procedure is carried out in logical steps, as it is when setting up any illumination system.

The illumination requirements of the phase-contrast microscope are fairly critical, and only a good quality, high-intensity lamp can be used. First, set up the microscope for Köhler illumination conditions with no annulus present in the condenser. Take care to get the illumination train as straight as possible. Focus the specimen and move the annulus appropriate for the phase objective into position on the condenser. Remove the eyepiece and insert the small telescope supplied with the phase-contrast kit in its place.

When this telescope is focused, the back focal plane of the objective can be seen with the dark phase ring in a central position. The position of this ring cannot be adjusted. Superimposed on this, and almost certainly out of position, is the bright image of the condenser annulus. Adjust the position of the image with the annulus centring screws, and focus it with the condenser focus control. Centre this image until it is exactly superimposed over the dark phase ring. Do not centre the image of the substage annulus with the condenser-centring controls, because the annulus has controls of its own.

If the image is unevenly illuminated, the lamp is not properly centred, and it must be adjusted until even illumination is obtained. Slight adjustments to the condenser height will have considerable effects on the brightness of the illumination. If the annulus image is misshapen and does not exactly fit the phase ring, there are alignment errors between the substage condenser and the objective lens. In these circumstances, it will be necessary to re-centre the condenser to the objective. This movement will also upset the annulus setting.

When alignment is complete, remove the telescope and replace the eyepiece. The microscope is then ready for use. Note that it is necessary to adjust each individual substage annulus to line up with its corresponding objective. On some

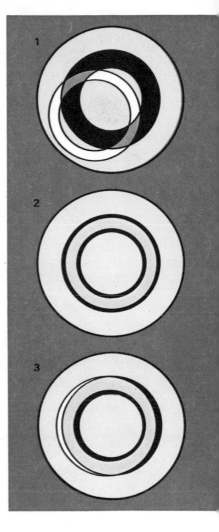

By using a special telescope in place of the eyepiece, the back focal plane of the objective can be viewed and the condenser annulus and objective phase-retardation plate aligned. 1 Annulus grossly displaced. 2 Annulus centred with adjusting screws on mounting. 3 Check the alignment after each change of annulus and objective.

microscopes there is individual adjustment. But on others, adjusting one upsets the rest, and the procedure must therefore be carried out each time an objective is changed. In practice, this is a less laborious operation than it sounds.

Uses of the microscope

The phase-contrast microscope has now become established as a routine laboratory instrument, even though it did not come into general use until after 1945. The microscope has particular uses in the study of living cells grown in tissue culture. This science would probably not have progressed so far without the use of phase-contrast techniques. An important early piece of work by K. Michel drew the attention of many biologists to phase-contrast. Using the microscope,

A living amoeba under phase contrast

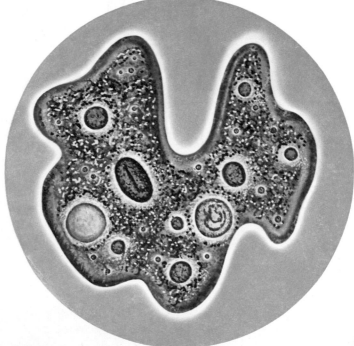

Michel studied the development of grasshopper sperm cells by making a time-lapse cine-film of the cells dividing.

Phase-contrast microscopy is not limited to the study of biological materials. It is also used in the study of any transparent material, including fibres, crystals, and plastics. It is not even restricted to the examination of transparent materials. Metals, for example, can now be examined by special phase-contrast techniques using reflected light.

The use of phase-contrast microscopy undoubtedly brings about a slight loss of resolution, but so much more can be seen under the phase-contrast microscope that this loss can easily be tolerated. The halo seen around images in the phase-contrast microscope arises because some of the diffracted light inevitably passes through the phase plate.

Cells from inside human cheek.
(*Top*) Ordinary light. (*Bottom*) Phase contrast

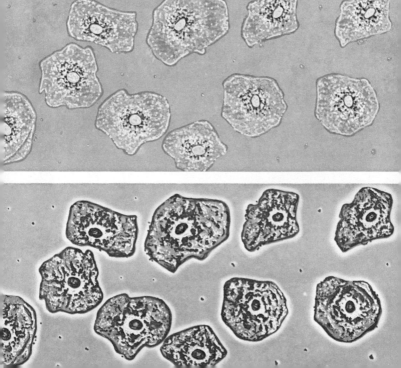

An interference microscope

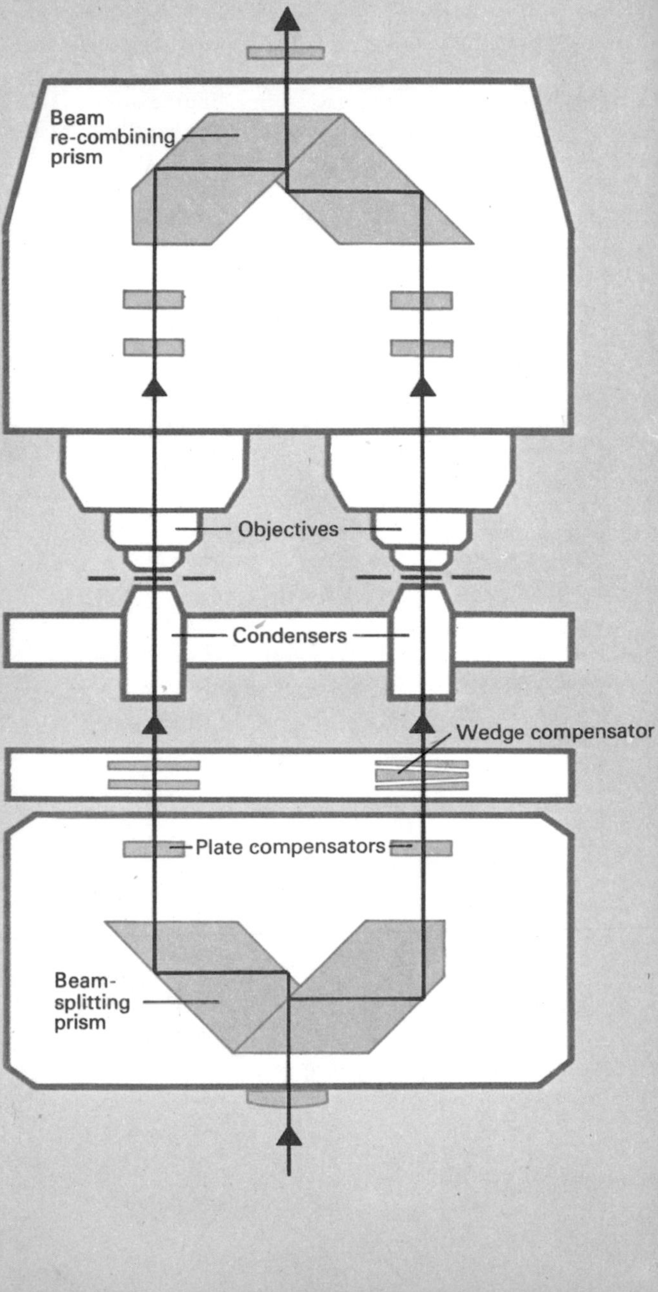

Beam
re-combining
prism

Objectives

Condensers

Wedge compensator

Plate compensators

Beam-
splitting
prism

The interference microscope

The *interference* microscope is a development of the phase-contrast microscope. In the interference microscope, two beams of light are used – one passes through the specimen and the other, called the *reference beam*, does not. The beam passing through the specimen is deviated as in the phase-contrast microscope, but the reference beam is entirely unaffected. The interference microscope consists, in effect, of two microscopes in one, each very accurately matched to the other except for the presence of the specimen in one. When the two beams are brought together at the top of the microscope, destructive interference takes place.

The most important feature of the interference microscope is that the path difference between the two beams produced by the specimen can be accurately measured. From these measurements, the refractive index of any particle in the specimen can be calculated and its mass determined. Also, this microscope does not suffer from the aberrations of the phase-contrast microscope. The image is seen clearly without the familiar halo which surrounds the image in the phase-contrast microscope. The interference microscope is also more sensitive to changes in the optical-path difference. As a result, smaller particles within the specimen can be examined, and areas of the specimen where the refractive index is only very slightly different from the surrounding medium can be clearly seen.

Another interesting feature of the interference microscope is that objects appear coloured when white light is used for illumination. This is because the interference affects only some of the wavelengths. Parts of the specimen that would probably go unnoticed under phase-contrast show up quite clearly in the interference microscope because the eye is very sensitive to different colours.

Both the phase-contrast microscope and the interference microscope are very valuable tools in the study of living biological systems. Whole new fields of study have been opened up by the introduction of these methods. This is particularly true now that the electron microscope, with its vastly superior resolution, has relieved the pressure from the optical microscope in the study of the most minute detail.

The fluorescence microscope

A very modern development of optical microscopy is the field of fluorescence microscopy. The fluorescence microscope is used to detect chemical bonding between certain substances. The reaction is made visible under the microscope by treating one of the substances in the reaction with a fluorescent dye.

Antigen-antibody reactions

A most important function of the fluorescence microscope is in the detection of what are known as *antigen-antibody* reactions. When a person or animal is infected with a disease, antibodies to that disease are produced and found in the blood. The animal then often becomes immune to the disease. The serum taken from such an animal is called an *immune serum* or *antiserum*. The antibodies produced are entirely specific for the disease which caused them to be produced. When this antiserum is mixed with its specific antigen – that is, the disease-causing agent – a reaction will take place, and the antibodies will become attached to the antigen.

These facts can be of great importance in the detection of disease. When a patient is suffering from an infection, his blood will contain specific antibodies to that infection. To treat the patient it is often necessary to find out which antibodies are present. To do this, some infected tissue from the patient is mixed with a series of known antisera. The one specific to the disease from which he is suffering will react, and an antigen-antibody reaction will take place. The problem that remains is how to detect the reaction, and for this the fluorescence microscope is used.

Serum antibodies are proteins, which can be 'labelled' by chemically combining them with fluorescent dyes. This process is called *conjugation*. Under ultra-violet or near ultra-violet light, these fluorescent dyes will be excited to produce visual colours. This is therefore a method of detecting the presence of antibodies. An antibody dyed in this way can also be used to detect when it enters into combination with an antigen in an antigen-antibody reaction. The reaction must be detected under the microscope because antibodies are, of course, extremely small.

These antigen-antibody reactions are not confined to infections with bacteria. Indeed, the presence of almost any foreign substance in an animal body will produce such a reaction. A rabbit can, for instance, be immunized against the red blood cells of a sheep. In the field of *immunology*, the study of immunity, new antigenic relationships are being discovered almost daily and the fluorescence microscope plays an important part.

A fluorescence microscope

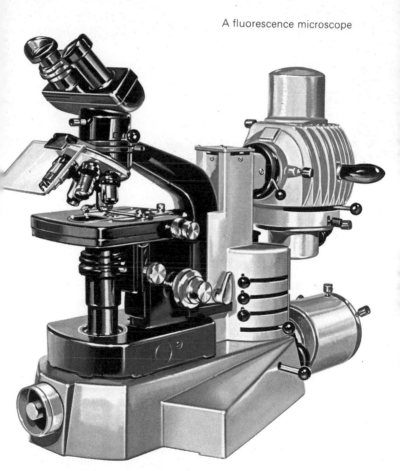

Fluorescence microscopy uses the blue-violet part of the spectrum.

How the microscope works

The most important feature of the fluorescence microscope is the illumination source, which must produce light rich in the ultra-violet and blue-violet parts of the spectrum. Light of these wavelengths excites the *fluorochrome,* or fluorescing part of the dye, to produce fluorescent light. A filter in the eyepiece blocks the ultra-violet light but allows the fluorescent light of a longer wavelength to pass. The points of fluorescent light therefore appear against a black background.

Ordinary tungsten lamps do not produce sufficient light of the correct wavelength, and so another source of illumination must be used. The most common source is the high-pressure, mercury-discharge lamp. These lamps get very hot and must be cooled either by forced draught from a fan or by water. They must also only be used with a protective cover, which prevents the very intense illumination and the ultra-violet radiation from reaching the operator and reduces the risk of accident if the globe should explode.

The most important part of the illuminator is the filter system. Filters are used in front of the lamp to let through to the specimen only the correct wavelength for the particular fluorochrome in use. They must cut off any visible light emitted by the lamp.

The lamp and microscope condensers must be able to transmit ultra-violet light. For this reason, they are made either of quartz or of a special crown glass which transmits ultra-violet light. Quartz condensers are better than glass because they are more resistant to cracking under the intense heat of the lamp. Once the illumination has been passed through the specimen the use of a special glass is neither necessary nor desirable.

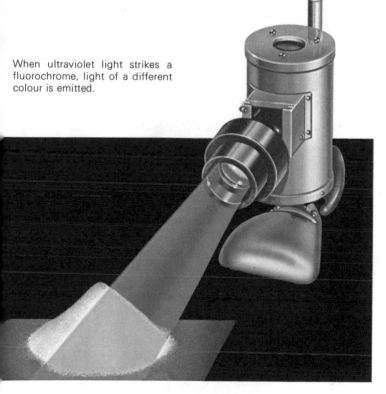

When ultraviolet light strikes a fluorochrome, light of a different colour is emitted.

In its simplest form, therefore, the fluorescence microscope consists of the following: a light source rich in ultra-violet; heat-absorbing filter; lamp condenser and diaphragm; primary ultra-violet filters; surface reflecting mirror; condenser; specimen; objective; eyepiece; and secondary filter designed to cut out any remaining ultra-violet light.

Iodine-quartz lamps can be used for a limited form of fluorescence microscopy, although they are more suitable for blue-light rather than ultra-violet fluorescence. The colour reactions produced in the fluorescence microscope are usually quite faint, and it is often necessary to use the microscope in dark conditions to see them at all. Photography with 35 mm. colour film is an essential part of this technique, although very long exposures are always required. Most fluorescence preparations fade rapidly and colour changes occur within minutes if specimens are left exposed to the illumination.

In fluorescence microscopy the objects appear coloured against a dark background.

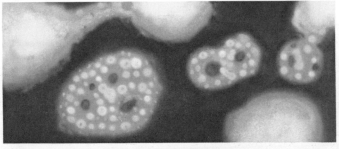

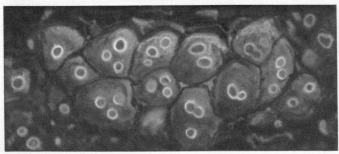

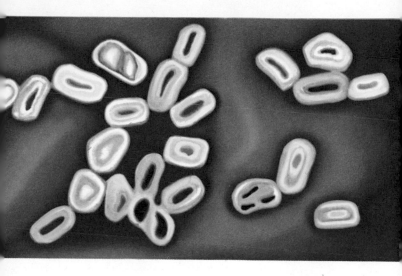

Bacteria under the fluorescence microscope

Uses of the microscope

Fluorescence microscopy has played an important rôle in investigations concerning many aspects of biological research. The fluorescence microscope is essential in the early diagnosis of cancer, because the technique requires examination of the microscopic distribution of specific antigens. In bacteriology, it plays an important part in the detection of bacteria as well as antigens. It has the great advantage over more traditional methods of being extremely rapid and accurate in laboratory use.

Many methods for the identification of hundreds of tissue components are now routine practice and the techniques have spread to other subjects such as the examination of botanical materials and the study of viruses.

Fluorescence microscopy has been developed only during the past twenty years as a routine laboratory technique. The growth of its use is an illustration of the fact that, even if the light microscope has reached its ultimate performance, new techniques and new uses for known techniques are still to be discovered.

SPECIAL MICROSCOPE TECHNIQUES

Measuring with the microscope

The measurement of both large and small objects with the microscope is common practice and a relatively simple procedure. Special microscopes which travel horizontally along a framework are used to measure large objects with great accuracy. But most microscopes are used for measuring very tiny objects.

The measurement of microscopic objects requires only two fairly inexpensive accessories – an eyepiece graticule and a stage micrometer. A special focusing eyepiece is an advantage but is not strictly necessary. The eyepiece graticule is a round disc of glass with a scale engraved on it which is inserted in the eyepiece at the level of the field stop.

To insert the graticule in the special focusing eyepiece, unscrew the bottom lens and remove the retaining ring or clip. Insert the graticule in the eyepiece with the scale uppermost and replace the clip and bottom lens. For an ordinary eyepiece, remove the top lens and drop the graticule down into the eyepiece on to the field stop. Then replace the top lens. The advantage of the focusing eyepiece is that the scale can be focused sharply. With the other system, some people may find that the scale is slightly blurred.

The next step is to place the eyepiece in the microscope tube and to focus the microscope on the scale which is engraved on the stage micrometer. This produces a field in which the scale of the eyepiece graticule is superimposed on the scale of the stage micrometer. We know the size of the stage micrometer scale, and so we can calculate the size that each division of the eyepiece scale represents. Many different stage micrometer scales are available. The most common one is graduated in one-hundredths of a millimetre. We can then use the eyepiece scale as a rule to measure the size of any microscopic detail of a specimen.

Usually the relationship of divisions between the stage micrometer and eyepiece graticule is most awkward. This can be overcome if the microscope has an adjustable tube-length for varying the magnification. We can set the two scales to a convenient figure by pulling the draw-tube in or

A micrometer is inserted in the eyepiece.

The divisions of the eyepiece micrometer are calibrated against a known micrometer placed on the stage.

Objects can now be measured using the eyepiece micrometer alone.

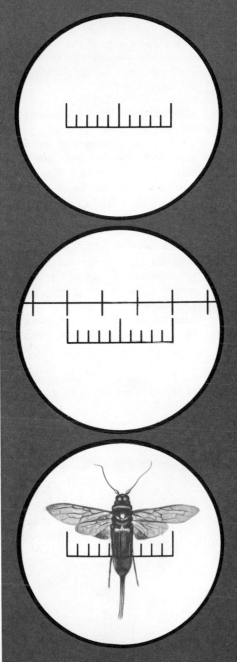

out. Make a note of the chosen draw-tube setting so that the draw-tube can always be set at this figure for measuring purposes. A separate calculation must, of course, be made for each objective.

Counting with the microscope

All the different types of eyepiece graticule available are used for measuring and counting purposes. For example, some have squared grids, some have round opaque dots and some have circles of different sizes. Many are made for a particular purpose, but in each case the stage micrometer is used for the initial calibration.

The microscope can be used to count small particles or objects, and the count can be expressed either as a percentage or as a count of actual numbers. Percentage counts are simply made by moving systematically from field to field, counting the numbers of each different object until a sum total of one hundred has been reached. Individual results are then expressed as a percentage of the total. It is often helpful when carrying out this sort of count to cut down the size of the field of view with a blanking square, which is cut from cardboard and inserted into the eyepiece at the level of the field stop.

The counting of total numbers of particles in suspension in liquid – for example, blood cells – is considerably more

Eyepiece blanking squares

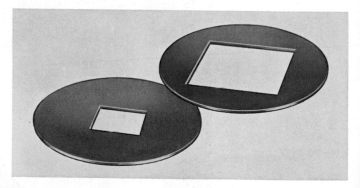

complicated and requires the use of a microscope *counting chamber*. The counting chamber consists of a thick microscope slide which has a trough of known depth ground into it. A special cover-slip fits across the trough. Engraved on the bottom of the trough is a grid area of known dimensions, which can be seen when the microscope is focused on it. The volume above the grid area of the counting chamber can therefore be calculated.

We fill the counting-chamber trough with the fluid containing the particles, and leave the particles to settle. We then focus the microscope on to the grid area and count the number of particles settled there. We know the volume of liquid from which these particles settled out, and we can therefore calculate the number present in each cubic millimetre, or cubic inch, of fluid. If the fluid contains large numbers of particles, as in the case of blood, we must dilute it until we obtain a suitable distribution of particles over the grid area. We must then, of course, take this dilution factor into account when making the final calculation.

The errors of counting by this technique are quite large, and it requires considerable standardization of method to get reproducible results.

A counting chamber

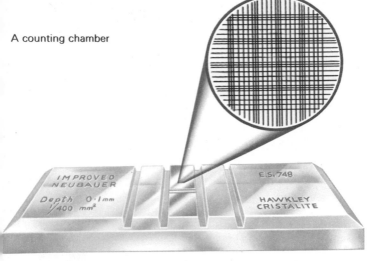

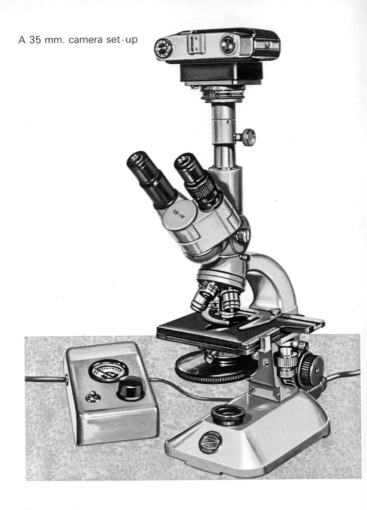

A 35 mm. camera set-up

Photography with the microscope

Photomicrography is the recording of microscope images by photographic methods. There are many ways of doing this, some of which can be done quite satisfactorily with very simple equipment. Strict Köhler illumination conditions are absolutely essential in photomicrography, and it is easiest to set up the microscope without the camera in place.

An eyepiece should always be used when taking micrographs. Without an eyepiece, the results are not satisfactory. The eyepiece is used to project the image on to the film in the camera. As a result, there is usually some magnification or reduction factor to take into account, depending on the distance from the film plane to the eyepiece. The elimination of field curvature is a big problem in photomicrography. For this reason, it is normal to take only the centre area of the field of view to eliminate the edges, where possible. Special eyepieces called *projection* or *photographic* eyepieces produce a slightly flatter field. In many, a mask inside them allows only the centre of the field to be projected.

Simple cameras

Without doubt, the simplest method of taking photographs down the microscope is by using a 35 mm. single-lens reflex camera fitted with a microscope adapter. Most camera manufacturers market such an adapter, which screws into the position normally occupied by the camera lens. In spite of the shortcomings of this system, quite reasonable results can be obtained.

One advantage of the single-lens reflex camera is that microscope focusing can be done by observation through the camera viewfinder. A disadvantage is that the camera is firmly fixed to the microscope, and shake caused by the shutter going off is transmitted to the microscope.

Exposure

Exposure setting is always a problem with a simple camera set-up. Special exposure meters for microscopes are usually very expensive, and normal exposure meters are not satisfactory. The usual method is to take test strips at various exposures and to select the exposure that gives the best results. The drawback to this method is that it is almost impossible to repeat exactly the diaphragm settings of the microscope each time it is set up. Taking test strips each time the camera is used is much too tedious.

However, the exposure latitude of black-and-white film is quite wide, and an experienced operator can judge exposure quite accurately. This is impossible with colour films. Also,

A vertical plate camera

black-and-white film is cheap enough for several exposures of each shot to be made. One disadvantage of 35 mm. film is that it is rather too small for photomicrography. A considerable amount of enlargement is required if prints of a reasonable size are to be obtained. For this reason, it is always necessary to use the finest-grain film, which is invariably the slowest.

However, the low speed of the film is of little consequence in microscope photography because long exposures are more desirable than short exposures. If the exposure is short (less than one second), the shake caused by the shutter opening and closing will occur for most of the exposure. But if the exposure is long, the shake will be for only a fraction of the exposed time.

Filters
The use of colour filters in photomicrography is a less complex subject that it may first seem. There are two main reasons for using filters – one is to increase or reduce contrast in the specimen and the other is to provide fairly monochromatic light. By using monochromatic light, we can eliminate

colour effects caused by the colour-sensitivity of the photographic emulsion, particularly when an ordinary achromatic objective is being used.

Another reason for using filters is to reduce the wavelength of the light in an effort to improve resolution. This is not entirely satisfactory, and for most purposes it can be ignored. Also, it is not necessary to use filters to correct colour aberrations when apochromatic objectives and panchromatic emulsions are being used.

Filters are therefore used mainly to increase or reduce contrast in a specimen. A specimen stained with haematoxylin and eosin will have bluish-black nuclei and reddish cytoplasm. The use of a red filter increases the contrast of the nuclei and reduces that of the cytoplasm. The use of a blue or bluish-green filter increases the contrast of the cytoplasm and reduces that of the nuclei. Therefore, if we use a filter which is the same colour as the specimen, there will be less contrast in the micrograph. But if we use a complementary filter, the contrast will be increased. For colour film, filters are not normally used, but it is essential to have a lamp burning at the correct colour temperature for the film and apochromatic objectives.

An optical bench used for photomicrography

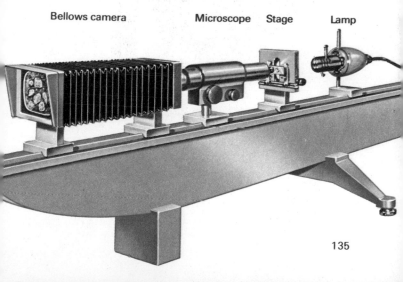

Bellows camera   Microscope   Stage   Lamp

135

Developing

The choice of developer for the film depends very much on personal preference, although fine-grain developers are almost always used. The loss of contrast resulting from the use of fine-grain developers is an advantage rather than a disadvantage because there is too much contrast in the photomicrographs of most stained specimens.

Additional equipment

More advanced microscope-camera arrangements incorporate beam-splitters so that focusing can be done with a special viewing eyepiece. Most of these eyepieces have focusing cross-hairs and view an aerial image. The advantage is that

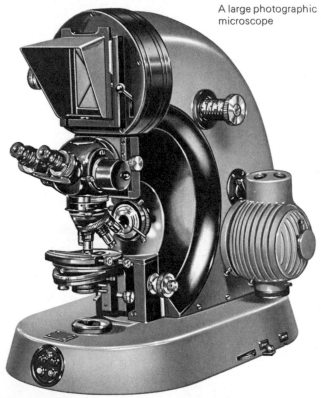

A large photographic microscope

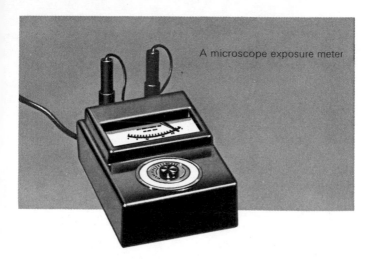

A microscope exposure meter

this image is bright and free from the interruption caused by ground glass and other devices, such as rangefinders, normally found in ordinary cameras. The disadvantage of these systems is that they are very expensive. Also, the presence of other optical components, such as beam-splitting prisms, in the optical path almost always reduces the quality of the final image.

Bellows cameras

For the best results, micrographs should be taken on photographic plates or cut-film using a large, bellows-extension camera. These cameras are very simple in construction and consist of a shutter mechanism connected to a plate-holder by an extending bellows. The whole system is either upright or horizontal, when all the components are usually housed on an optical bench. Usually the lower end of the camera is not in physical contact with the microscope. A light-tight sleeve from the camera fits over, but just clears, the microscope tube.

The microscope lamp is often placed so that it shines directly up the microscope, eliminating the need for the mirror. The simplest arrangement of lamp, condenser, specimen, objective, eyepiece and photographic plate will produce the best results. Focusing is carried out on a ground-glass screen fitted in the position that the plate will occupy.

The bellows system gives continuously variable magnification and should be calibrated so that reproducible results can be otained. Normally, apochromatic lenses will be used in the system, and the bellows will be extended so that only the middle of the field is photographed. In fact, better results are usually obtained by using a lower-power objective and longer bellows extension than by using a high-power objective with short bellows extension. There is usually a clear area in the centre of the ground-glass focusing screen on which an exposure meter or eyepiece can be placed. With an eyepiece, more critical focus can be obtained.

Specialized microscopes

We have so far considered only the more conventional types of transmission microscopes. But there are many other, more specialized microscopes which can normally be used only for a specific purpose. A few of them are mentioned below.

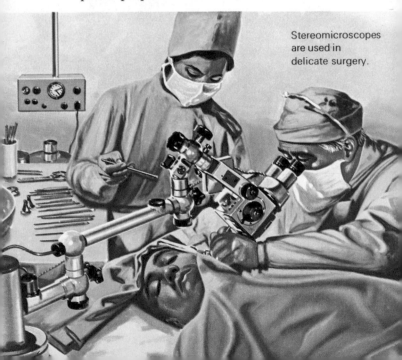

Stereomicroscopes are used in delicate surgery.

In metallurgy, the microscope is adapted for the examination of completely opaque specimens. In the *metallurgical* microscope, semi-silvered mirrors or prisms are used to direct the illumination down the objective lens from a source half-way up the body-tube. The light strikes the polished surface of the specimen and is reflected back up the microscope. In this way, the surface of the metal specimen can be examined at high magnification, and almost the full resolving power of the microscope is retained.

Simple metallurgical microscope

A hot stage microscope

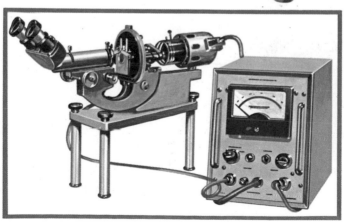

Some microscopes are designed so that dynamic experiments can be carried out and viewed in microscopic detail at the same time. Specimens can be examined while they are stressed, heated, and cooled. Even experiments into wear caused by rubbing surfaces can be observed. In a *hot-stage* microscope, the specimen is attached to a thermocouple, which both heats it and measures its temperature.

Special microscopes are used by surgeons performing delicate operations. They must have a long working distance and a wide field of view. The operator focuses and moves the microscope by foot controls, leaving his hands free. Powerful illumination is directed from a point between two stereoscopic objectives on to the working area.

Some microscopes are inverted, with the stage on top and the objectives and viewing system below. The eyepiece position is corrected by using prisms so that normal viewing is achieved. With a microscope like this, cells growing in a culture medium can be observed without being disturbed.

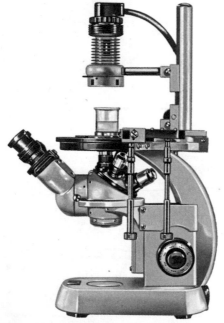

An inverted microscope

Geologists use *petrological* microscopes, which incorporate two polaroid filters, to observe how the specimen alters the plane of the light rays. The results are very colourful.

Biologists often find it useful to place the microscope in a warm box to keep it and its immediate environment at body temperature. By this means living specimens, which normally die if they are cooled, can be observed for long periods.

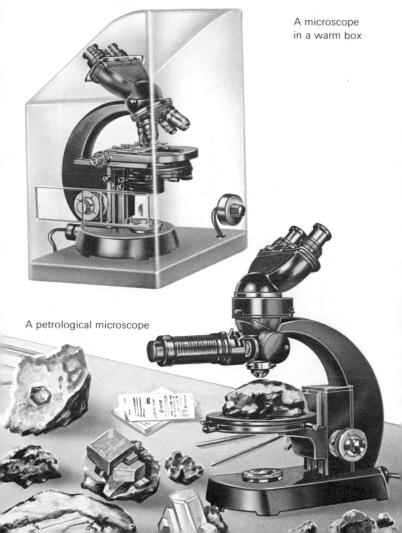

A microscope in a warm box

A petrological microscope

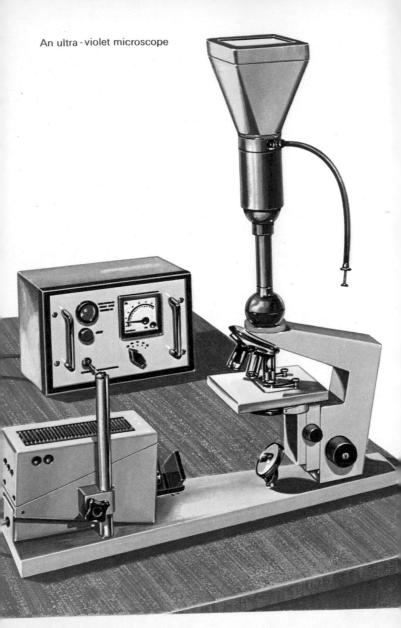

An ultra - violet microscope

The ultra-violet microscope

The most severe drawback of the light microscope is its relatively poor resolving power. The work of Abbe has shown that the resolving power (R) can be represented by the following equation:

$$R = \frac{0 \cdot 6 \; \lambda}{N.A.}$$

where λ is the wavelength of the illumination and N.A. is the numerical aperture (see page 39).

Microscope lenses with numerical apertures close to the theoretical maximum have been available since the late 1800s. Therefore the only way to achieve an improvement in resolution is to shorten the wavelength of the light used. For this purpose, Köhler and von Rhohr developed the ultra-violet microscope in 1904. They hoped that it would achieve a resolution twice as good as the conventional microscope.

Unfortunately, there are many problems involved in ultra-violet microscopy. The human eye is not sensitive to light of this wavelength and therefore another viewing system must be used to see and focus the image. A fluorescent screen can be used to convert the image into a visible one, but the resolving power of the screen is so poor that, once focused, the image must be recorded photographically.

Normal lens glass will not transmit ultra-violet light, and the lenses in the ultra-violet microscope must therefore be made either of fluorite or quartz. The slide on which the specimen is mounted must also be made of quartz for the same reason. To further complicate matters, most media used to mount specimens are opaque to ultra-violet light.

Mirrors, or reflecting objectives, are used in the more highly developed ultra-violet microscopes to overcome the problem of ultra-violet transmission. They consist of a system of mirrors instead of lenses. They have the added advantage that focusing can be carried out in ordinary white light, and ultra-violet light substituted for the photographic exposure. This cannot be done with ordinary lenses because there is a change of focus with change of wavelength.

Ultra-violet microscopy is very complicated, and the results barely justify its complexity. The ultra-violet microscope has become almost obsolete since the introduction of the electron microscope.

THE ELECTRON MICROSCOPE

History and development

In 1879, J. J. Thomson, working at the Cavendish Laboratory in Cambridge, showed that rays emitted from the cathode of electrodes sealed in an evacuated glass tube were composed of streams of negatively charged particles, which he termed *electrons*. Work on these new rays followed in all the major physics research centres of Europe. In 1924, L. de Broglie showed that these negatively charged particles could be considered to travel in a wave form, like light.

In 1926, Busch showed that rays of electrons could be focused by surrounding them with a magnetic coil, and that an electron 'lens' could be made, although he himself never made one. In the following year Gabor produced the first electron lens which consisted of a solenoid cased in soft iron. It was realized that the electron lens could form the basis of an 'electron' microscope which would have a very high resolving power, because electrons have a very short wavelength – only fractions of an *Angstrom unit* (one millionth of a millimetre).

The first electron microscope was made by M. Knoll and E. Ruska in Berlin in 1928. It had only two lenses and magnified only 17 times. By 1931 the magnification had been extended to 400 times, although it was still not possible to put specimens into the microscope. In those early days, it was thought that a practical electron microscope might never be achieved. It was difficult to maintain the high vacuum that is needed inside the microscope column (electrons cannot travel readily through air). Also, it was felt that this vacuum and the electron bombardment would ruin any specimens that were used.

However, the work continued despite these difficulties. By 1933, Ruska had built a microscope which had an accelerating voltage for the electrons of 75,000 volts. It could accept specimens and had water-cooled lenses. Pictures were obtained on a fluorescent screen at the bottom of the microscope column. The resolving power of this microscope was five times that of the best light microscope. The first pictures of biological material were made in 1934. It was obvious

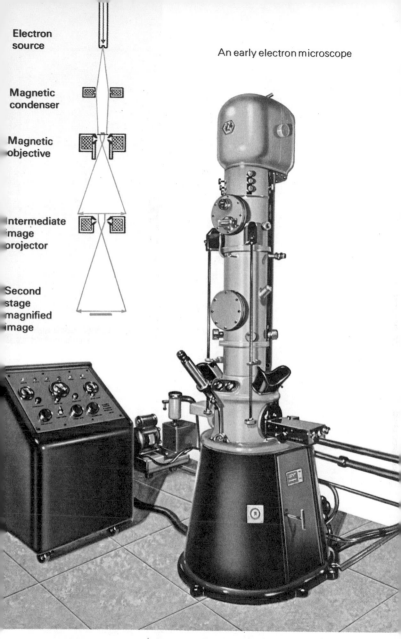

Electron source

Magnetic condenser

Magnetic objective

Intermediate image projector

Second stage magnified image

An early electron microscope

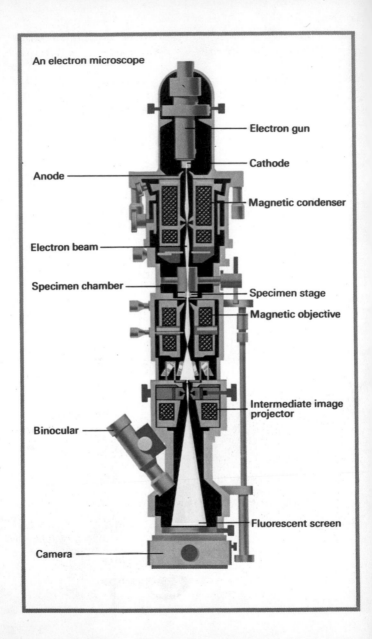

An electron microscope

Electron gun

Cathode

Anode

Magnetic condenser

Electron beam

Specimen chamber

Specimen stage

Magnetic objective

Intermediate image projector

Binocular

Fluorescent screen

Camera

even at this early stage that preparation of the specimen was going to be a limiting factor in the performance of the electron microscope.

Development was naturally slow between 1939 and 1945, but since then it has proceeded at such a rapid pace that resolving powers in the region of two Angstrom units are now commonplace. The resolution obtainable still depends on the specimen and not on the microscope.

How the electron microscope works

The general layout of the illumination system and lenses of the electron microscope corresponds to the layout of the light microscope. The electron 'gun' which produces the electrons is equivalent to the light source of the optical microscope. The electrons are accelerated by a high-voltage potential (usually 40,000 to 100,000 volts) and pass through a condenser lens system usually composed of two magnetic lenses. This system concentrates the beam on to the specimen, and the objective lens provides the primary magnification. The final image in the electron microscope must be projected on to a phosphor-coated screen so that it can be seen. For this reason, the lenses that are the equivalent of the eyepiece in an optical microscope are called *projector* lenses.

Normally, the electron microscope is upside-down when compared with the light microscope, with the electron gun at the top of the column and the fluorescent screen at the bottom. The screen is viewed through a window let into the front. The column of the microscope is held under high vacuum to prevent the electrons passing through it from striking air molecules and being scattered.

The strength of an electron lens depends on the current passing through the coil which produces the magnetic field. The strength of the lens can be varied by altering the current. In the electron microscope, therefore, the lenses are fixed, and adjustments are made to magnification and focus by altering the current passing through the lens coils. The condenser lens focuses the beam of electrons on to the specimen and affects the amount of illumination on the screen; the objective lens focuses the image; and the projector lenses alter the magnification.

Lens defects

Magnetic lenses suffer from the same defects (chromatic and spherical aberration) in the same way as glass lenses. But the same methods of correction cannot be used, because there is no 'negative' electron lens.

A very small lens aperture is employed to correct spherical aberration, but this severely limits the final resolution. Chromatic aberration is reduced by using electrons of a single wavelength. To produce such electrons, the accelerating voltage must be kept very steady because the wavelength of the beam is related to the accelerating voltage.

Electron-microscope lenses suffer from the further aberration of *astigmatism*, which affects light-microscope lenses to a far lesser degree. Astigmatism is caused by the lens having two focal planes for axes at right angles to each other. Nothing can be done about astigmatism in an optical microscope. But in the electron microscope, it can be corrected. Astigmatism in the electron microscope arises from two sources – from the lenses themselves and from dirty apertures in the objective lens, which are only 25 to 50 microns in diameter. In both cases, the astigmatism can be corrected by a skilled operator. In effect, the electron microscope achieves its superior resolution more in spite of, than because of, its lenses.

Shown here greatly enlarged, the electron-microscope grid which supports the specimen is only 3 mm. in diameter.

The equipment

Most electron microscopes consist of two major units – a *desk unit* and a *power unit,* each of which may weigh up to half a ton. The desk unit has the column attached to it and houses all the controls for the lenses, vacuum, and so on, together with the vacuum pumps. The power unit supplies the high voltage used to accelerate the electrons and the very stable current to the lens. The electron microscope must be installed in a room which is reasonably free from vibrations and stray magnetic fields.

The column is fitted with air-locks so that the specimen and camera plates or film can be changed without losing the whole of the vacuum in the column. Changing a specimen usually takes about five minutes.

Most electron microscopes are equipped with a binocular light microscope, which magnifies about 10 times and is used to view the fluorescent screen where the image is formed. To see resolution in the region of five Angstrom units, it is necessary to use magnifications of millions of times. High-resolution electron microscopes have a top-screen magnification in the region of 250,000 times.

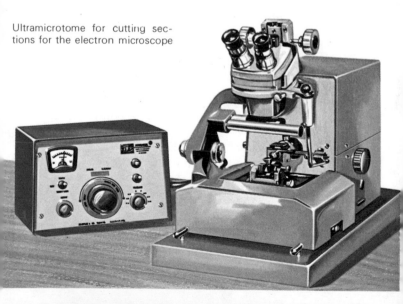

Ultramicrotome for cutting sections for the electron microscope

Photographs are always taken for critical examination of specimens. The camera is situated below the fluorescent screen, which is raised to expose the plate. Photographic enlargement of the plate gives a print with the desired final magnification. The resolution of the photographic plate is far superior to that of the fluorescent screen, and detail which is not visible on the screen may well be visible in the finished micrograph.

Specimen preparation

Preparing specimens for the electron microscope is an extremely critical operation. Electrons have a very poor penetrating power, and specimens must therefore be extremely thin. Increasing the accelerating voltage of the electrons improves penetration but results in loss of contrast in the image. Sections are prepared in a similar manner to those for the light microscope, except that embedding is done not in wax but in a plastic material such as methacrylate or epoxide resin. Only very small pieces of tissue (about a cubic millimetre) can be processed, but this is of little consequence when the magnification of the microscope is considered.

Sectioning is done under a stereoscopic dissecting microscope. A glass or diamond knife on an ultra-microtome cuts sections about one-twentieth of a micron thick. The sections are floated directly from the cutting edge on to a trough of water attached to the knife.

The sections are picked up on a small copper grid, about 2 to 3 mm. in diameter. The sections are viewed through the holes in the grid.

It is often necessary to provide some kind of support film for the specimen to prevent it from falling through the holes and to make it more stable in the electron beam. The most stable support film is carbon. The film is prepared by striking an arc across two carbon electrodes in a high vacuum. The carbon evaporated from the electrodes is collected on a clean micro-slide, floated on to water, and picked up on the grid. The only disadvantage with using carbon films to support the specimen is that the film structure can be seen in modern high-resolution microscopes and adds undesirable background to the micrographs.

A modern electron microscope

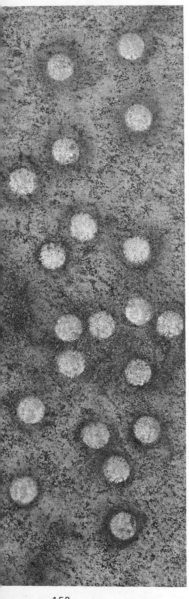

Staining

Sections for the electron microscope can be 'stained' to increase contrast. The 'stains' used are electron-dense (electron absorbing) substances, such as lead or other heavy metals. 'Staining' is carried out either by soaking the piece of tissue in a solution of the heavy metal before sectioning, or by floating the grid with sections attached on to a similar solution.

Particles in suspension, such as viruses, are usually mixed with an electron-dense material and sprayed on to carbon-coated grids. In the microscope, the particles can be seen against the dark electron-dense background. This form of negative 'staining' has been of immense value in the study of virus particles.

Shadow casting

Another interesting technique in electron microscopy is that of 'shadow' casting to give the appearance of a third dimension in the micrographs. Particles in suspension are sprayed on to a grid and allowed to dry. Then a

Electron micrograph of virus particles

metal, such as gold, is evaporated under vacuum on to the grid, which is held at an angle. When viewed in the microscope, the particles appear to cast shadows. The shadows are pale compared with the rest of the background, which has received a coating of gold. The familiar '3D' electron micrographs, with the shadows dark, can be produced by reversing the photographic negative. This technique is of limited value because the coating of metal obscures detail.

Uses of the electron microscope

The electron microscope is not limited to the examination of the sub-microscopic anatomy of specimens. Biochemists, for example, use it to check that the substances they are analyzing are the correct ones. When, for example, the amount of DNA present in mitochondria is being analyzed, the cells are broken up and centrifuged at high speed. The resulting pellet is then sectioned and examined in the electron microscope to check that the fraction being

Electron micrograph of muscle fibres

used for analysis is the correct one. Techniques such as these have brought about great advances in our knowledge of the cell and the functions of its various components.

It is, of course, impossible to study living cells in the electron microscope because of the high vacuum to which they are subjected and also because the electron beam itself is highly damaging to living tissue. Some attempts have been made to study living cells using a special stage in the microscope, but the results achieved so far have been very discouraging. The fact that electron micrographs show a picture of biological material that has been grossly altered by fixation and subsequent processing makes the interpretation of results very difficult. But since these results are constantly repeatable, this situation must be of considerable significance.

Future developments

The electron microscope is still at an early stage of development. Its resolving power is gradually being improved, and it is probable that one day it will be able to resolve 0·1 Angstrom. The greatest problem at the moment is that the techniques of specimen preparation are unable to keep pace with the development of the microscope. However, this was the case with the light microscope not so many years ago, and it is now possible to achieve the full light-microscope resolution with almost any well-prepared specimen. The problems of repeating this achievement for the electron microscope are unquestionably much more complex. However, it should not be very long before cell structure can be seen at a sub-molecular level.

Electron microscopes that have a million volts available to accelerate the electrons are now being built, mainly for the metallurgist. These microscopes can be used to examine much thicker specimens than the conventional electron microscope can, and there is evidence that they may have the added advantage of causing less beam damage to the specimen. Perhaps microscopes such as these may enable us to see the living cell molecules at work. We are without doubt at the beginning of another era in microscopy.

The million-volt microscope

BOOKS TO READ

For general introductions to microscopes and microscopical techniques, the following books are recommended and are usually available from bookshops and public libraries.

Animal Tissue Techniques by G. L. Humason. W. H. Freeman, London and San Francisco, 1962.

The Evolution of the Microscope by S. Bradbury. Pergamon Press, London, 1967.

An Introduction to Medical Laboratory Technology by F. J. Baker, R. E. Silverton, E. D. Luckcock. Butterworth, London, 1966.

Modern Microscopy by V. E. Cosslett. G. Bell and Sons, London, 1966.

Teach Yourself Microscopy by W. G. Hartley. English Universities Press, London, 1962.

Using the Microscope by A. L. E. Barron. Chapman and Hall, London, 1965.

INDEX

159

SOME OTHER TITLES IN THIS SERIES

Natural History

The Animal Kingdom
Australian Animals
Bird Behaviour
Birds of Prey
Fishes of the World
Fossil Man
A Guide to the Seashore

Life in the Sea
Mammals of the World
Natural History Collecting
The Plant Kingdom
Prehistoric Animals
Snakes of the World
Wild Cats

Gardening

Chrysanthemums
Garden Flowers

Garden Shrubs
Roses

Popular Science

Atomic Energy
Computers at Work
Electronics

Mathematics
The Weather

Arts

Architecture
Jewellery

Porcelain
Victoriana

General Information

Flags
Military Uniforms
Rockets & Missiles
Sailing

Sailing Ships & Sailing Craft
Sea Fishing
Trains

Domestic Animals and Pets

Budgerigars
Cats
Dogs

Horses & Ponies
Pets for Children

Domestic Science

Flower Arranging

History & Mythology

Discovery of
  Africa
  North America
  The American West
  Japan

Myths & Legends of
  Ancient Egypt
  Ancient Greece
  The South Seas